Hydrogen Technologies - Advances, Insights, and Applications

Edited by Zak Abdallah and Nada Aldoumani

Published in London, United Kingdom

Hydrogen Technologies - Advances, Insights, and Applications
http://dx.doi.org/10.5772/intechopen.1001641
Edited by Zak Abdallah and Nada Aldoumani

Contributors
Abir Ben Abdallah, Akbar Dauletbay, Beibei Wang, Carine Alves, Gustavo Castro, Hao Shen, Kheireddin Kadri, Luciano Hocevar, Malcolm Penman, Rodrigo Coelho, Steven Snow, Sébastien Ballut, Trisha Hoover, Xiaodan Wang

First published in London, United Kingdom, 2024 by IntechOpen
IntechOpen is the global imprint of INTECHOPEN LIMITED, registered in England and Wales, registration number: 11086078, 167-169 Great Portland Street, London, W1W 5PF, United Kingdom

British Library Cataloguing-in-Publication Data
A catalogue record for this book is available from the British Library

Additional hard and PDF copies can be obtained from orders@intechopen.com

Hydrogen Technologies - Advances, Insights, and Applications
Edited by Zak Abdallah and Nada Aldoumani
p. cm.
Print ISBN 978-1-83769-604-8
Online ISBN 978-1-83769-603-1
eBook (PDF) ISBN 978-1-83769-605-5

For EU product safety concerns:
IN TECH d.o.o., Prolaz Marije Krucifikse Kozulić 3, 51000 Rijeka, Croatia,
info@intechopen.com or visit our website at intechopen.com.

Meet the editors

Dr. Zak Abdallah is currently a senior lecturer in sustainability at the University of the West of England (UWE), with a focus on advanced materials and hydrogen technology. He is the principal investigator (PI) of an ongoing EPSRC project on digital investment casting of metallic structures, collaborating with the Advanced Manufacturing Research Centre (AMRC) and the University of Sheffield, UK. At UWE, Dr. Abdallah is also developing expertise in additive manufacturing of polymeric, composite, and metallic structures, supporting a research group comprising students and researchers. He is an associate at the Centre for Print Research (CFPR), where he supervises multiple research projects related to additive manufacturing. The CFPR at UWE is known for its excellence in additive manufacturing, utilizing advanced robotic-driven approaches. Before joining UWE, Dr. Abdallah was the principal and lead academic of fracture and fatigue at the Steel and Metals Institute (SaMI) at Swansea University. In this role, he was involved in tribology, surface engineering, metal joining, and hydrogen-induced corrosion. Dr. Abdallah has also worked at the Rolls-Royce University Technology Centre (UTC) at Swansea University on materials-related projects for gas turbine applications, including nickel, titanium, and steel alloys. Additionally, he has served as a materials consultant at Swansea Materials Research and Testing (SMaRT), working on various industrial projects for companies such as Airbus, McLaren, Rolls-Royce, and others.

Dr. Nada Aldoumani is the director of research at ITECH Engineering, a registered consultancy firm in the UK with headquarters in London. This firm provides services to industrial partners, including training and modeling activities related to industrial operations. Dr. Aldoumani specializes in services related to metallic and composite structures, drawing from her extensive knowledge in materials science and structural engineering. Her consultancy supports customers worldwide, offering expertise in the design of modern materials and engineered applications, with a particular focus on finite element (FE) modeling and MATLAB coding. Dr. Aldoumani is a computational engineering scientist with a Ph.D. in composite materials uncertainty and modeling, and over nine years of experience working with UK industries and higher education institutions. She is an expert in composite materials modeling using ANSYS, and her work is supported by numerous publications in internationally recognized peer-reviewed journals on uncertainty quantification and FE modeling of materials. She is also a co-editor of five internationally published scientific books in materials science and modeling, and has a strong background in structural engineering, particularly in indeterminate structures. Previously, Dr. Aldoumani was a fellow researcher and lecturer at Swansea University, where she pioneered the field of uncertainty quantification in composite structures. During her tenure, she developed a novel approach that automates ANSYS through MATLAB code, capable of running thousands of trials to obtain probabilistic Gaussian distributions.

Her research has been published in internationally recognised peer-reviewed journals. Additionally, she has supervised undergraduate and postgraduate students in composite materials testing and modeling, and has worked as a consultant in fatigue, creep, and fracture mechanics of materials.

Contents

Preface

This book delves into hydrogen technology as an alternative fuel option for future engineering applications. Hydrogen is widely regarded as the leading candidate to replace fossil fuels, outperforming electric battery vehicles. It is a clean energy source that produces zero emissions during use, making it highly appealing for future applications. However, hydrogen is not entirely green if produced using fossil fuel energy; thus, employing renewable energy sources for its production enhances its environmental benefits throughout its life cycle. Electrolysis is considered the most effective method for hydrogen production, despite its high energy demands. There are significant challenges associated with the storage and transportation of hydrogen, necessitating further research to develop innovative materials and processes to address these issues. This book includes chapters that explore various approaches to hydrogen production, as well as the negative impacts of hydrogen on materials during transportation and storage. It serves as a valuable reference for researchers worldwide who are focused on hydrogen technology, a highly relevant topic today. With my expertise in materials science and hydrogen technology, alongside the co-editors' expertise in materials and modeling, we have successfully compiled this book. We are proud to present it to the research community and are optimistic about its contribution to a greener environment in the future. We hope you enjoy reading the book and find it a rich source of information to further research in this engineering field.

Dr. Zak Abdallah
Senior Lecturer (Materials and Hydrogen Technology),
University of the West of England (UWE),
Bristol, UK

Dr. Nada Aldoumani
Director of Research,
ITECH Engineering Consultancy,
London, UK

Section 1

Hydrogen Technology

Chapter 1

Hydrogen Technologies: Recent Advances, New Perspectives, and Applications

Carine Alves, Gustavo Castro, Rodrigo Coelho and Luciano Hocevar

Abstract

Hydrogen has become a crucial element in the search for clean energy solutions. It provides promise as a versatile and sustainable energy carrier. This chapter discusses the history of hydrogen technologies, tracing its journey from early industrial uses to its current pivotal role in modern energy systems. It explores the versatility and energy storage capabilities of hydrogen, emphasizing its potential for decarbonization in various sectors such as transport, industry, and electricity generation. The chapter aims to provide a comprehensive overview of recent advancements in hydrogen technologies, examining innovative production methods and groundbreaking applications. Through this exploration, it seeks to clarify the role of hydrogen in shaping global energy landscapes and contributing to sustainable practices. By assessing its impact across different sectors, the chapter highlights the significance of hydrogen in promoting a transition toward cleaner and more resilient energy systems that align with environmental goals and the pursuit of carbon neutrality.

Keywords: hydrogen, new perspectives, energy, renewable energy, biofuels

1. Introduction

The energy sector finds itself at a pivotal juncture, marked by escalating demand and pressing global environmental concerns [1]. Hydrogen stands as a pivotal clean energy carrier in the modern energy landscape, offering a sustainable solution to global energy challenges. Its significance extends beyond being merely an energy source; it embodies a potential keystone for a low-carbon future. Historically, hydrogen technologies have evolved significantly, transitioning from early industrial uses to a cornerstone of contemporary clean energy systems.

Initially utilized in chemical processes and refining, hydrogen's role has expanded due to its versatility, energy storage capabilities, and potential for decarbonization. Its ability to integrate with various energy sectors, including transport, industrial processes, and electricity, positions hydrogen as a crucial player in the transition toward sustainable energy practices. The historical context of hydrogen technologies reflects a journey of innovation and adaptation, underscoring its growing importance in achieving a balanced and resilient energy ecosystem that aligns with environmental goals and the pursuit of carbon neutrality [2].

In Europe, ambitious plans are underway to achieve climate neutrality by 2050, as outlined in the Paris Agreement [3]. Social imperatives underscore initiatives like the European Green Deal and its transformative "Fit for 55" package, aimed at driving decarbonization within the EU. Geopolitical dynamics further shape the energy landscape, evident in initiatives such as the "REPowerEU" plan, which seeks to swiftly reduce reliance on Russian fossil fuels [4].

Despite significant strides in renewable energy adoption, the intermittency of renewables poses a persistent challenge to their widespread integration into the energy mix. Addressing this hurdle, hydrogen (H_2) has emerged as a compelling solution for leveraging surplus electricity generated by sources like solar and wind parks, thereby enhancing the flexibility of energy systems [5]. Widely acknowledged as a key player in the green energy transition, hydrogen holds immense potential as an energy carrier with diverse applications across industries [6].

While currently utilized primarily in industrial processes and derived from fossil fuels, hydrogen is poised for a transformative expansion into sectors like transportation and construction in the coming decades, increasingly produced as a low-carbon fuel [7]. Countries worldwide have outlined roadmaps and national strategies for fostering the hydrogen economy, with deployment targets set for the next decade and beyond [8].

This transition heralds not just technological advancements but also significant cultural shifts, necessitating the development of robust market infrastructure and widespread accessibility to hydrogen resources. The evolution of the hydrogen economy is influenced by a myriad of factors, reflecting a diverse array of approaches across various disciplines [9]. In evaluating its sustainability, economic and environmental considerations often take center stage, comparing hydrogen technologies against conventional fossil fuels using metrics such as cost and carbon footprint.

However, it is crucial to recognize that prevailing global hydrogen energy models tend to prioritize techno-economic parameters, potentially diverging from societal preferences. Thus, while efficiency and cost-effectiveness remain vital, achieving a truly sustainable hydrogen economy demands a holistic approach that aligns with broader societal aspirations and values.

2. Recent advances in hydrogen production

This section delves into the latest advancements in hydrogen production, a critical pillar for the hydrogen economy's viability and sustainability. It explores cutting-edge developments across various hydrogen production technologies, emphasizing innovations that enhance efficiency, reduce costs, and minimize environmental impacts. The focus is on breakthroughs in electrolysis, photoelectrochemical, and biological hydrogen production methods, each contributing uniquely to the diversification and resilience of clean hydrogen supply. By examining these recent advancements, the section aims to provide a thorough understanding of the current state-of-the-art technologies and their potential to revolutionize the hydrogen production landscape, paving the way for a sustainable energy future.

2.1 Electrolysis

Electrolysis, a principal method for hydrogen production, has undergone significant technological transformations, enhancing its appeal for large-scale, clean hydrogen generation. Recent years have seen remarkable progress in the development of advanced electrode materials and catalysts. These innovations have been pivotal in increasing the efficiency and durability of electrolyzers, particularly in proton exchange membrane (PEM) electrolysis (**Figure 1**). Enhanced catalysts offer reduced overpotentials, leading to lower energy consumption and increased operational efficiency. The advent of novel, earth-abundant catalysts also promises to reduce costs and reliance on scarce materials like platinum and iridium, making electrolysis more economically viable and sustainable.

Modern electrolysis systems are being designed with a focus on integration with renewable energy sources, such as solar and wind power. This integration is crucial for producing green hydrogen, ensuring that the hydrogen generation process is entirely carbon neutral. Innovations include adaptive systems that can operate efficiently under variable power inputs, enhancing the compatibility of electrolyzers with fluctuating renewable energy outputs, thereby optimizing the overall efficiency and cost-effectiveness of hydrogen production (**Figure 2**).

High-temperature electrolysis (HTE) has gained attention for its potential to improve the overall system efficiency. By operating at elevated temperatures, HTE can utilize heat (often waste heat from industrial processes) to reduce the electrical energy required for hydrogen production. Recent advancements have focused on improving the durability and performance of high-temperature electrolyzers,

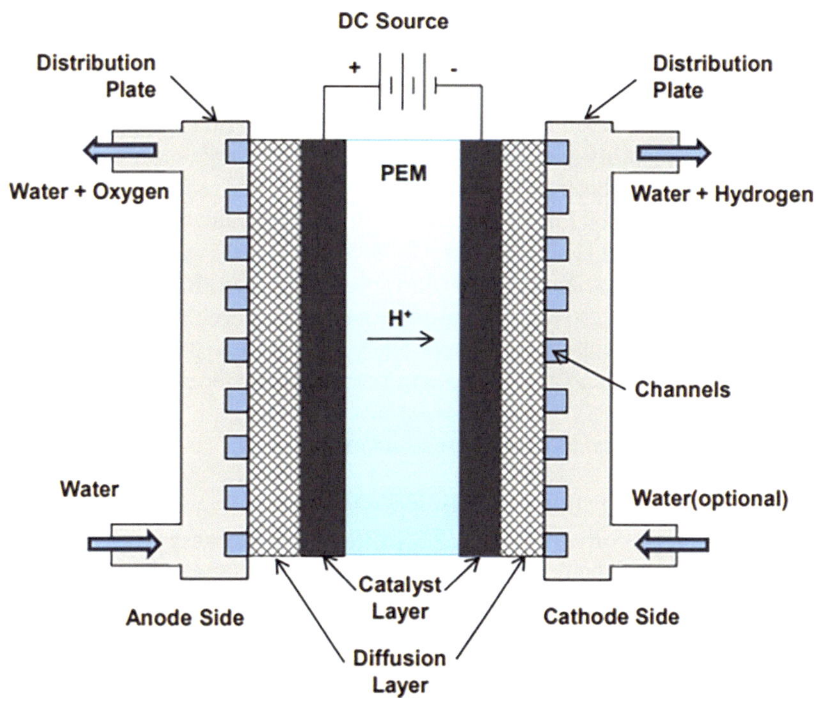

Figure 1.
PEM electrolyzer cell [10].

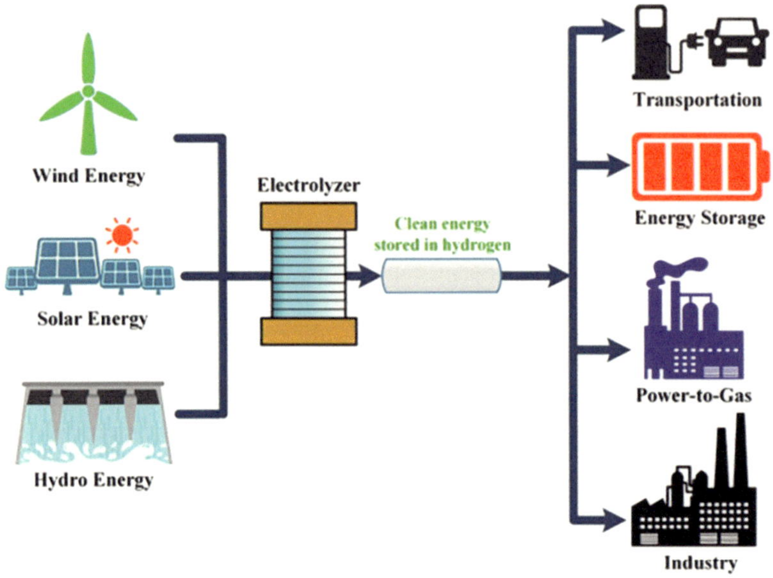

Figure 2.
Modern electrolysis systems [11].

developing robust solid oxide electrolyte materials that can withstand harsh operating conditions while maintaining high efficiency. High-temperature steam electrolysis, also known as steam electrolysis, is another promising method to produce hydrogen using the thermal energy produced by nuclear reactors. Unlike conventional water electrolysis, HTSE utilizes higher temperatures ranging from 800°C to 1000°C to split water molecules, thus requiring less electricity. This makes HTSE a more efficient and cost-effective method for hydrogen production [12].

The push toward commercializing electrolysis technology has led to significant advancements in scalability and the economic feasibility of large-scale operations. Efforts to scale up electrolyzer manufacturing, reduce capital costs, and improve system longevity are crucial for widespread adoption. The industry is moving toward gigawatt-scale production facilities, driven by decreasing costs and increasing demand for green hydrogen, especially in sectors like transportation, industry, and grid storage. These advancements collectively contribute to the ongoing evolution of electrolysis as a sustainable and economically feasible technology for hydrogen production, aligning with global energy transition goals and the growing emphasis on decarbonization.

2.2 Photoelectrochemical hydrogen production

Photoelectrochemical (PEC) hydrogen production represents a cutting-edge approach that directly converts solar energy into chemical energy, offering a promising pathway to sustainable hydrogen generation. PEC systems utilize semiconductor materials to absorb sunlight and generate electron-hole pairs, which then drive water-splitting reactions to produce hydrogen. The efficiency of these systems depends heavily on the semiconductor's ability to absorb solar radiation, facilitate charge separation, and catalyze water reduction and oxidation reactions. Recent innovations focus on optimizing these materials for better light absorption, increased charge carrier mobility, and enhanced catalytic properties (**Figure 3**) [13].

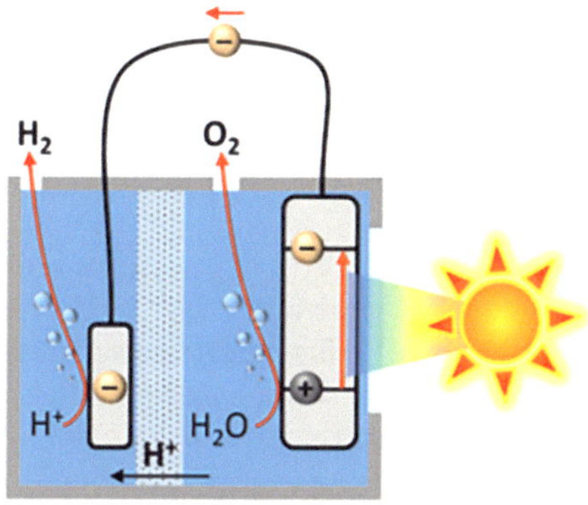

Figure 3.
PEC system [13].

The core of the PEC technology lies in the development of efficient and stable semiconductor photocatalysts. Recent research has concentrated on novel materials like perovskites, quantum dots, and nanostructured semiconductors, which offer superior light absorption and charge separation efficiencies. Innovations in material engineering, such as doping, heterojunction formation, and surface passivation, have significantly enhanced the photocatalytic performance and stability under solar irradiation.

The design of PEC cells has evolved from traditional planar configurations to more advanced designs that maximize light absorption and minimize recombination losses. Concentrating PEC cells, tandem cell structures, and integrated PEC systems that combine photovoltaic cells with electrolyzers have emerged, showing enhanced solar-to-hydrogen conversion efficiencies. These innovative architectures are crucial for optimizing the spatial arrangement of materials and the overall integration of components within the PEC system.

Integrating PEC systems with existing renewable energy infrastructures offers a pathway to decentralized hydrogen production. Such integration enables the utilization of excess renewable energy to produce hydrogen, aligning hydrogen production with periods of high renewable availability. This synergy can enhance the overall efficiency of renewable energy systems, providing a storage solution in the form of hydrogen and facilitating a more resilient energy grid.

Despite significant progress, PEC hydrogen production faces challenges related to efficiency, durability, and scalability. Efficiency improvements are needed to make the technology economically competitive, while durability issues must be addressed to ensure long-term operation without significant degradation. Research is focused on developing protective coatings, advanced water-splitting catalysts, and engineered interfaces to overcome these challenges and improve the long-term stability of PEC cells.

The future of PEC hydrogen production is promising, with ongoing research aimed at overcoming current limitations and moving toward practical applications. The field is advancing toward the development of cost-effective, scalable, and efficient PEC systems capable of operating under real-world conditions. Collaborations between academia, industry, and government entities are vital in accelerating the translation of laboratory-scale innovations to commercial-scale applications, paving the way for PEC technology to contribute significantly to the global hydrogen economy. These insights reflect the dynamic nature of PEC hydrogen production research, highlighting the technological advancements and the potential of this method in contributing to a sustainable energy future.

2.3 Biological hydrogen production

Biological hydrogen production is an area of growing interest due to its potential for sustainable and eco-friendly fuel generation. This section outlines the significant advancements and insights in this field. Biological hydrogen production primarily involves two pathways: biophotolysis of water by algae and photofermentation by bacteria (**Figure 4**). These processes harness the natural ability of microorganisms to produce hydrogen from water or organic substrates, using sunlight or metabolic energy. Recent research has focused on enhancing these biological systems' efficiency, scalability, and reliability, leveraging genetic engineering and biotechnological innovations.

Advances in genetic engineering have enabled the optimization of microbial strains for increased hydrogen yield and productivity. Scientists have manipulated the metabolic pathways of microorganisms like cyanobacteria and green algae to enhance their hydrogen-producing capabilities, focusing on the overexpression of hydrogenase enzymes and the suppression of competing metabolic processes that consume hydrogen. Biological hydrogen production is a challenging area in biotechnology, with environmental and energy source problems. However, in the past decade, hydrogen energy has made progress in all areas of energy. High potential exists in existing technologies for the development of practical H_2 production bioprocesses. Increasing rates of synthesis and final yields of H_2 require further research and development. Bioprocess integration, optimization of bioreactor design, rapid removal, purification of hydrogen, directed evolution of hydrogenase, and metabolic engineering of the H_2-evolving microorganism offer exciting prospects for biohydrogen systems. Novel strategies will also be very encouraging and exciting in the future. The rapid advances in biological and engineering sciences will facilitate overcoming existing bottlenecks and new challenges and create new opportunities for economical hydrogen production shortly [14].

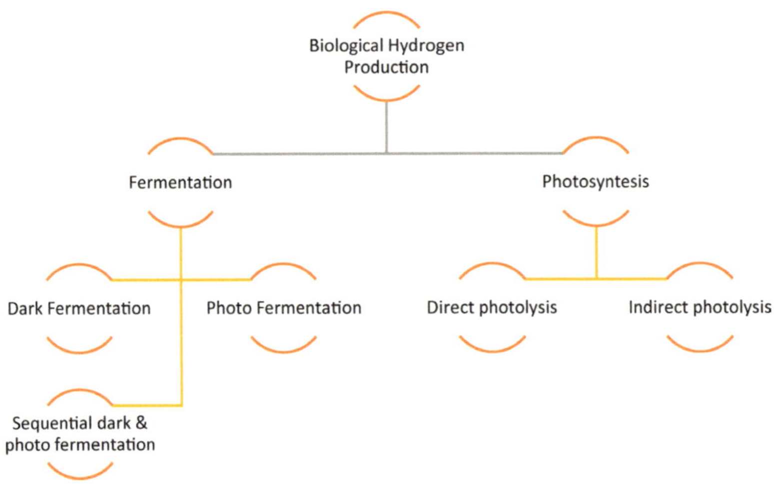

Figure 4.
Biological hydrogen production methods.

The development of hybrid systems combining biotechnological and chemical processes represents a significant leap forward. These systems aim to optimize hydrogen production by integrating different biological and physicochemical stages, allowing for the sequential degradation of organic compounds and subsequent hydrogen generation, thereby increasing the overall process efficiency and yield. Innovations in bioreactor design have been crucial for scaling up biological hydrogen production. High-performance bioreactors with optimized conditions for microbial growth and hydrogen production have been developed, including photobioreactors for algae and fermenters for anaerobic bacteria. These systems are designed to maximize light penetration, nutrient supply, and waste removal, enhancing the efficiency and sustainability of hydrogen production.

Despite promising laboratory-scale results, scaling up biological hydrogen production to commercial levels poses significant challenges. These include maintaining stable microbial cultures over long periods, optimizing the production process for high-volume output, and ensuring economic viability. Ongoing research is addressing these challenges by developing robust microbial strains, efficient bioreactor systems, and integrated process technologies. The future of biological hydrogen production is promising, with the potential to contribute significantly to the renewable energy mix. The field is moving toward more sustainable and cost-effective solutions, focusing on utilizing waste biomass as a feedstock and integrating production systems with existing waste treatment facilities. This approach not only produces clean energy but also contributes to waste reduction and resource recovery, aligning with circular economy principles. The exploration of biological hydrogen production is a testament to the innovative approaches being pursued in the renewable energy sector, offering a glimpse into a future where clean fuel generation harmonizes with environmental stewardship.

3. Advances in hydrogen storage technologies

This section addresses the critical advancements in hydrogen storage technologies, a vital component for the widespread adoption of hydrogen as a clean energy carrier. It explores innovative solutions and recent breakthroughs that have significantly improved the efficiency, safety, and economic viability of hydrogen storage. From high-density solid-state systems to liquid organic carriers and advanced composite materials, this section will delve into the latest developments that are shaping the future of hydrogen storage, highlighting their potential to transform energy systems and facilitate a seamless transition to a hydrogen-based economy.

3.1 Solid-state hydrogen storage

Solid-state hydrogen storage is a key technology that offers high hydrogen density and the potential for safe, efficient energy storage, crucial for the widespread adoption of hydrogen as a clean energy carrier. Advances in materials science have led to the development of new solid-state hydrogen storage materials, such as complex metal hydrides, chemical hydrogen storage materials, and intermetallic compounds. These materials can absorb and release hydrogen through chemical reactions or physical processes, offering high storage densities and the potential for tailored thermodynamics and kinetics.

Researchers [15] have made significant strides in understanding and improving the hydrogen storage mechanisms of solid-state materials. This includes optimizing

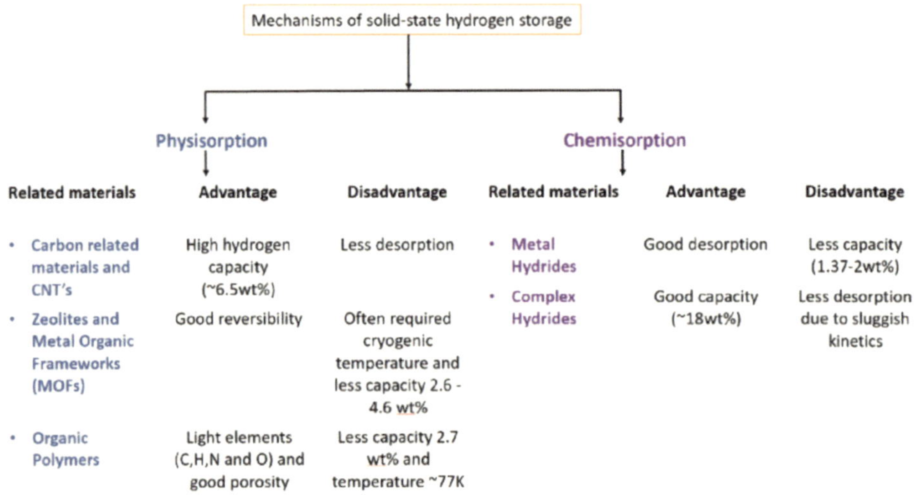

Figure 5.
Solid-state hydrogen storage mechanisms [15].

the absorption/desorption kinetics, increasing reversible hydrogen capacity, and reducing the operating temperatures required for efficient hydrogen release and uptake (**Figure 5**).

The integration of nanotechnology has been a game-changer in solid-state hydrogen storage, enabling the manipulation of materials at the atomic or molecular level to improve their hydrogen storage properties. Nanostructuring of storage materials can enhance surface area, modify reaction pathways, and improve kinetics, leading to faster charging and discharging rates and better overall performance. Innovations in system design and engineering have facilitated the development of more compact, efficient, and safer solid-state hydrogen storage systems. These systems are designed to optimize heat and mass transfer, accommodate the volumetric expansion of materials, and integrate seamlessly with fuel cell technologies and other hydrogen applications [16].

Despite progress, challenges remain in realizing the full potential of solid-state hydrogen storage, such as high activation energies, slow kinetics, and issues with scalability and cost. Ongoing research is focused on addressing these challenges through material modification, advanced engineering strategies, and the development of hybrid systems that combine the advantages of different storage methods. The future of solid-state hydrogen storage looks promising, with ongoing research pushing the boundaries of material performance and system integration. The aim is to achieve storage solutions that are not only efficient and safe but also economically viable for large-scale applications, paving the way for hydrogen to become a cornerstone of the global energy landscape. The advancements in solid-state hydrogen storage are pivotal for the transition to a hydrogen economy, offering solutions that could transform energy storage, transportation, and various industrial applications.

3.2 Liquid organic hydrogen carriers

Liquid organic hydrogen carriers (LOHCs) represent a transformative approach in hydrogen storage technologies, offering a safe, efficient, and energy-dense alternative to conventional storage methods. LOHCs are organic compounds capable of chemically binding and releasing hydrogen through hydrogenation and dehydrogenation reactions. They offer a practical solution for hydrogen storage and transportation, as they can operate at near-ambient conditions, making them safer and more adaptable to existing infrastructure compared to high-pressure or cryogenic hydrogen storage methods. The idea of storing hydrogen in LOHCs at room temperature is based on the process of hydrogenation and dehydrogenation of organic molecules (**Figure 6**). LOHCs are a type of hydrogen storage system that remains in a liquid state when it contains a high concentration of hydrogen. They are a potentially cost-effective, safe, and easily manageable way of storing hydrogen [17]. Additionally, LOHCs enable long-term energy storage without any losses or boil-off and can be transported easily. Key to the efficiency of LOHC systems is the development of advanced catalysts that facilitate rapid and reversible hydrogenation and dehydrogenation reactions.

Recent breakthroughs have focused on enhancing the activity, selectivity, and durability of these catalysts; reducing energy requirements; and improving the overall economic viability of the LOHC technology. Innovations in system design are crucial for optimizing the performance of LOHC systems. This includes the integration of advanced thermal management systems, improved reactor designs for enhanced kinetics, and scalable process configurations that can adapt to varying demands and operational conditions, ensuring efficient and consistent hydrogen release and storage [17]. The search for optimal LOHC materials has led to the exploration of various organic compounds, including aromatics, carbazoles, and heterocycles. These materials are evaluated based on their hydrogen storage capacity, stability, recyclability, and compatibility with existing infrastructure, driving the development of more effective and sustainable LOHC solutions.

Despite their potential, LOHC systems face challenges such as the need for high temperatures in dehydrogenation, the potential toxicity of certain carriers, and the long-term stability of the storage material. Research is directed toward overcoming these challenges by developing more efficient catalytic systems, identifying environmentally benign and cost-effective carriers, and enhancing the material's recyclability and lifecycle. The future of the LOHC technology is promising, with ongoing research aimed at commercial-scale applications, integration with renewable energy sources, and development of mobile and stationary hydrogen storage solutions. The adaptability of LOHC systems to existing liquid fuel infrastructure presents a unique opportunity to facilitate the widespread adoption of hydrogen as a key energy carrier in the transition toward a sustainable energy economy. These

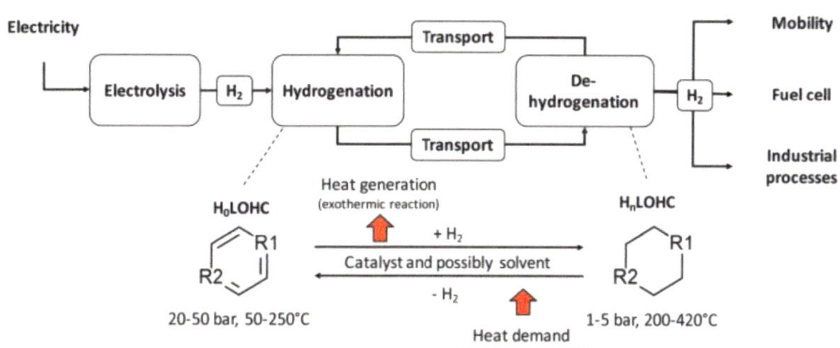

Figure 6.
LOHC system [17].

detailed insights into LOHCs underscore their potential to revolutionize hydrogen storage, offering a viable pathway for the safe, efficient, and large-scale deployment of hydrogen energy solutions.

3.3 Advanced composite materials

Advanced composite materials for hydrogen storage are at the forefront of innovation, offering unique solutions that combine lightweight properties, high hydrogen storage capacities, and excellent safety profiles. The development of composite materials involves the integration of various substances, such as metal hydrides, polymers, and carbon-based materials, to create systems that can store hydrogen with high efficiency. These composites often leverage the synergistic properties of their constituents, offering enhanced hydrogen storage capacities, improved kinetics, and reduced risks of leakage [18].

Recent advances in materials science and research have enabled engineers to explore the use of less conventional materials in their designs. Composite materials are becoming increasingly popular in marine, aerospace, automotive, and other industries and offer excellent strength-to-weight ratios, improved thermal and mechanical properties, and other desirable qualities that result from combining different constituent materials. The use of composites in hydrogen storage tanks could improve the efficiency of such systems [18–20].

Looking forward, the emphasis on sustainability is driving the development of composite materials that are not only efficient and safe but also environmentally benign. The potential for recycling and reusability of these materials is an important consideration, aiming to ensure that the hydrogen economy is supported by circular economy principles. The exploration of advanced composite materials is crucial for overcoming the current limitations of hydrogen storage, offering promising prospects for the development of safe, efficient, and sustainable hydrogen energy systems.

4. New perspectives in hydrogen utilization

This section explores the evolving landscape of hydrogen utilization, highlighting innovative applications and the integration of hydrogen technologies into various sectors. It examines the latest trends and developments that are expanding the role of hydrogen beyond conventional uses, focusing on its potential in energy systems, industrial processes, and emerging markets. The discussion will encompass recent advancements in hydrogen fuel cells, combustion technologies, and the role of hydrogen in decarbonizing heavy industries, underscoring its significance in achieving a sustainable, low-carbon future.

4.1 Hydrogen fuel cells

Hydrogen fuel cells are a transformative technology in the energy sector, offering a clean, efficient alternative to conventional combustion-based power generation. Hydrogen fuel cells generate electricity through the electrochemical reaction of hydrogen and oxygen, producing water as the only byproduct. Recent advancements have focused on enhancing the efficiency, durability, and cost-effectiveness of these cells, broadening their applicability in energy systems. Significant strides have been made in developing advanced materials for electrodes, electrolytes, and catalysts,

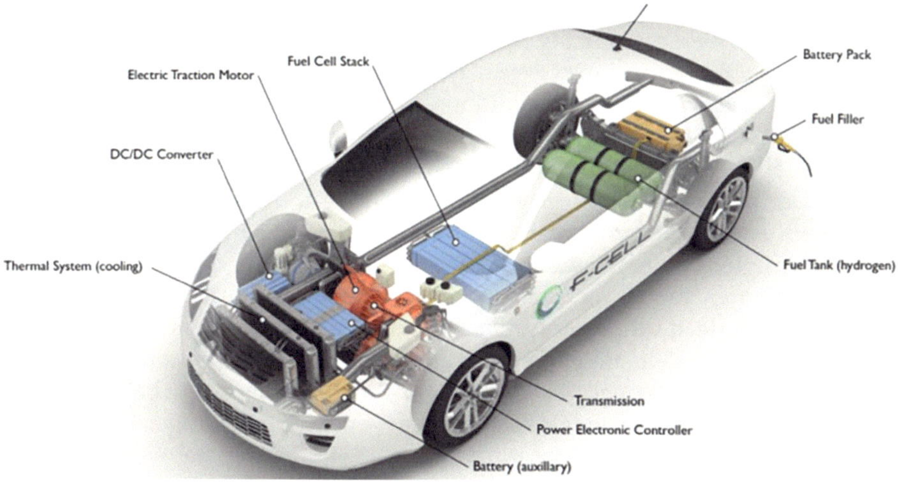

Figure 7.
Hydrogen fuel cell vehicles (FCVs) [21].

aiming to improve performance and reduce the costs of fuel cells. Innovations include the use of non-precious metal catalysts, durable polymer electrolytes, and improved electrode architectures that enhance the electrochemical surface area and catalytic activity.

Efforts in system integration involve optimizing fuel cell stacks for various scales and applications, from portable devices to large stationary installations. Scalability challenges are being addressed through modular design, allowing for the customization of power output to meet specific demands, and enhancing the versatility of fuel cell applications. The automotive industry has seen significant developments in hydrogen fuel cell vehicles (FCVs) (**Figure 7**), with improvements in fuel cell efficiency, vehicle range, and refueling infrastructure. These vehicles offer a sustainable alternative to traditional combustion engines, with the added benefits of quick refueling times and long driving ranges.

Beyond transportation, hydrogen fuel cells are gaining traction in stationary applications for backup power, grid support, and distributed energy systems, as well as in portable applications for consumer electronics, remote power, and emergency systems, showcasing their flexibility and adaptability across energy domains. While hydrogen fuel cells present a promising path toward decarbonization, challenges remain in terms of large-scale hydrogen production, infrastructure development, and market acceptance. Future research and development are geared toward addressing these challenges, with a focus on improving the economic viability and environmental impact of fuel cell technologies. The advancements in hydrogen fuel cells are pivotal for the transition to a cleaner energy landscape, offering innovative solutions for transportation, stationary power, and beyond, marking a significant step toward a sustainable, low-carbon future.

4.2 Hydrogen combustion

Hydrogen combustion is emerging as a pivotal technology in the transition to a cleaner energy future, offering new perspectives and applications. Hydrogen combustion is characterized by its high energy yield and clean emission profile, primarily producing water vapor when burned. The efficiency of hydrogen combustion engines is continually enhanced through innovations in combustion chamber design, fuel injection systems, and ignition timing optimization, aiming to maximize the energy output while minimizing any NOx emissions. Recent developments have focused on refining the combustion process to ensure complete hydrogen burn, reduce emissions, and improve overall system efficiency. Advancements in turbocharging, exhaust gas recirculation, and lean-burn technologies have been crucial in enhancing the performance and environmental compatibility of hydrogen combustion engines [22].

Hydrogen combustion is increasingly integrated with renewable energy sources. It serves as a flexible load for power systems, capable of storing excess renewable energy and releasing it when demand peaks. This integration facilitates a more resilient and sustainable energy infrastructure, leveraging hydrogen's potential to balance and store renewable energy. Beyond its role in power generation, hydrogen combustion is gaining traction in industrial applications, including high-temperature processes in steelmaking, glass production, and other manufacturing sectors. It offers a pathway to decarbonize industrial heat processes, significantly reducing the carbon footprint associated with these industries [23].

Despite its potential, challenges remain in the widespread adoption of hydrogen combustion, such as ensuring safety, improving public perception, and developing infrastructure. Innovations in safety technologies, public engagement strategies, and the expansion of hydrogen refueling infrastructure are critical to overcoming these barriers. The future of hydrogen combustion looks promising, with ongoing research aimed at enhancing efficiency, reducing costs, and expanding applications. As the technology matures and scales up, hydrogen combustion is set to play a crucial role in a diverse range of sectors, from transportation to industrial heating, contributing significantly to global decarbonization efforts. Hydrogen combustion stands as a transformative technology, poised to significantly impact how energy is produced and utilized, driving forward the clean energy transition with its high efficiency and potential for broad application.

4.3 Hydrogen in chemical synthesis

The integration of hydrogen into chemical synthesis signifies a transformative shift toward greener chemical processes, leveraging hydrogen's potential as a clean and sustainable reactant. Hydrogen is pivotal in developing eco-friendly synthesis routes, offering a cleaner alternative to traditional processes that rely on fossil fuels. Its application in chemical synthesis reduces carbon emissions, aligning with global sustainability goals and the increasing demand for green chemistry. This section elaborates on the multifaceted role of hydrogen in chemical synthesis [24].

Innovations in catalysis have revolutionized hydrogen's role in chemical synthesis, enabling more efficient, selective, and lower-energy processes. Catalytic systems that facilitate hydrogenation reactions have become more robust, versatile, and capable of driving reactions under milder conditions, thus enhancing their industrial applicability and environmental compatibility. In the realm of organic and inorganic synthesis, hydrogen is employed as a powerful reducing agent. Its ability to facilitate the reduction of various functional groups while maintaining a clean environmental profile is paramount in synthesizing a wide array of chemicals, pharmaceuticals, and materials. Leveraging hydrogen in chemical synthesis has led to process intensification, where reactions can be conducted more rapidly, efficiently, and at lower temperatures. This not only improves the economics of chemical production but also significantly reduces the energy footprint, contributing to more sustainable industrial practices [25].

The synthesis of green chemicals using hydrogen can be further enhanced by integrating the processes with renewable energy sources. Electrochemical hydrogen generation from water, powered by solar or wind energy, provides a sustainable pathway for producing hydrogen that can be directly utilized in chemical manufacturing, closing the loop on a truly green process [26].

The integration of hydrogen into chemical synthesis holds the promise of revolutionizing the chemical industry, making it cleaner and more sustainable. However, challenges such as the development of cost-effective, efficient, and durable systems for hydrogen production and utilization, as well as the adaptation of industrial infrastructure, need to be addressed to fully realize this potential. The utilization of hydrogen in chemical synthesis not only stands as a testament to the versatility and sustainability of hydrogen but also opens new avenues for the development of green manufacturing processes, marking a significant step toward a sustainable chemical industry.

5. Emerging applications of hydrogen technologies

This section delves into the groundbreaking applications of hydrogen technologies that are shaping the future of energy, industry, and transportation. It explores innovative uses that extend beyond conventional boundaries, highlighting how hydrogen is being integrated into new markets and sectors. From powering zero-emission vehicles to enabling large-scale renewable energy storage and supporting decarbonization in various industries, this exploration sheds light on the versatility of hydrogen and its potential to drive significant environmental and economic benefits.

5.1 Hydrogen in transportation

Hydrogen is increasingly recognized as a pivotal energy carrier in the transportation sector, offering a sustainable alternative to fossil fuels. Hydrogen fuel cells are a key technology in powering zero-emission vehicles, particularly in the automotive industry. They convert hydrogen into electricity, generating only water vapor and heat as byproducts, thus providing a clean alternative to internal combustion engines. Recent advancements in fuel cell technology have enhanced efficiency, reduced costs, and extended the lifespan of these systems. These improvements are crucial for the commercial viability of hydrogen-powered vehicles, including cars, busses, and trucks.

The expansion of hydrogen refueling infrastructure is critical to the adoption of hydrogen in transportation. Efforts are underway globally to increase the number of hydrogen refueling stations, which is vital for supporting the widespread use of fuel cell vehicles. Hydrogen is particularly suited for heavy-duty and long-range transport applications, such as trucks, busses, and maritime vessels, where battery-electric solutions face challenges due to weight and range limitations.

Hydrogen transportation can be integrated with renewable energy sources, providing a pathway to reduce greenhouse gas emissions significantly. Using green hydrogen, produced from water electrolysis powered by renewables, enhances the environmental benefits of hydrogen-powered transport. The shift to hydrogen-powered transportation has significant economic and environmental implications. It offers a pathway to reduce dependency on fossil fuels, decrease air pollution, and mitigate greenhouse gas emissions, contributing to the goals of the Paris Agreement and global climate change mitigation efforts. The integration of hydrogen technologies in transportation is set to revolutionize the sector, offering sustainable, efficient, and clean solutions that align with global energy transition and decarbonization objectives.

5.2 Hydrogen for grid balancing and energy storage

The role of hydrogen in grid balancing and energy storage is increasingly pivotal as the world shifts toward renewable energy sources. Hydrogen serves as a key enabler for the integration of intermittent renewable energy sources like wind and solar power. By converting excess electricity into hydrogen, it acts as an energy buffer, ensuring that surplus renewable energy is not wasted but stored for future use.

Hydrogen technologies offer grid balancing services by providing demand response capabilities. They can quickly release stored hydrogen to generate electricity during peak demand periods or when renewable energy supply is low, thereby maintaining grid stability and reliability. Unlike batteries that are typically used for short-term energy storage, hydrogen can be stored for long durations without significant losses. This characteristic makes it ideal for seasonal storage, enabling energy systems to overcome periods of prolonged energy surplus or deficit.

Hydrogen can be produced and stored at a local level, supporting the decentralization of energy systems. This local production and utilization of hydrogen contribute to reducing energy transmission losses and enhancing local energy resilience. Hydrogen plays a crucial role in sector coupling, linking the electricity sector with heating, transport, and industry. By converting electrical energy into hydrogen, it can be used across different sectors, promoting an integrated and flexible energy system.

The future of hydrogen in grid balancing and energy storage looks promising, with ongoing innovations aimed at improving the efficiency, scalability, and cost-effectiveness of hydrogen-based energy storage solutions. As renewable energy adoption grows, the importance of hydrogen in ensuring a stable, reliable, and sustainable energy grid becomes increasingly significant. Hydrogen's versatility and energy storage capacity position it as a key player in the transition to a more resilient, efficient, and sustainable energy system, particularly in the context of increasing global renewable energy penetration.

5.3 Hydrogen in industrial processes

The utilization of hydrogen in industrial processes (**Figure 8**) is a transformative approach, heralding a new era of sustainability and efficiency in various sectors. Hydrogen is instrumental in decarbonizing heavy industries like steel, cement, and chemicals. These sectors, traditionally dependent on fossil fuels, are now exploring hydrogen to reduce their carbon footprint, leveraging its clean combustion and high-energy content. Integrating hydrogen into industrial processes enhances efficiency, reduces waste, and promotes cleaner production methods. Hydrogen's role in processes like refining, ammonia production, and methanol synthesis is evolving, driven by the demand for greener practices.

In industries where high-temperature heat is essential, hydrogen is emerging as a key energy carrier. Its combustion releases water vapor, providing the necessary

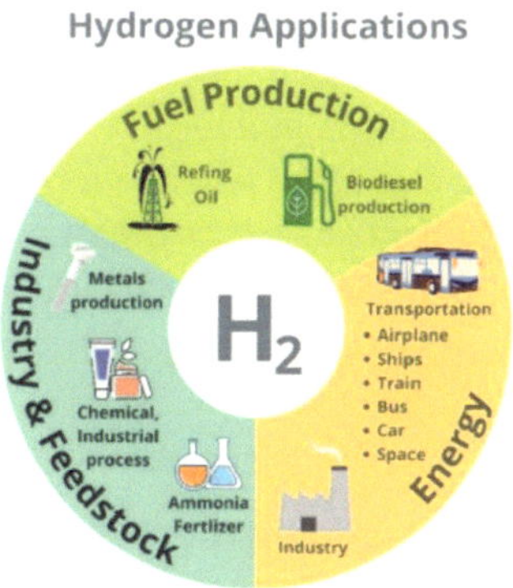

Figure 8.
Hydrogen applications [27].

energy without the carbon emissions associated with coal or natural gas. Research and development are paving the way for innovative applications of hydrogen in industries, including its use in advanced manufacturing, as a reducing agent in metallurgical processes, and in creating sustainable synthetic fuels. The industrial adoption of hydrogen is supported by evolving economic and policy frameworks that incentivize clean energy use. Investments in hydrogen infrastructure, subsidies for clean energy, and carbon pricing are pivotal in integrating hydrogen into industrial applications.

The future of hydrogen in industrial processes is linked closely with global sustainability goals. As industries aim to reduce their environmental impact, hydrogen stands out as a versatile and clean solution that aligns with the global shift toward renewable energy and circular economy principles. Hydrogen's integration into industrial processes is not just an innovation in energy use but a necessary step toward achieving a sustainable industrial future, reducing global carbon emissions, and facilitating the transition to a clean energy economy.

6. Conclusions

In this comprehensive exploration of hydrogen technologies, we delved into the latest advancements and innovative applications shaping the future of energy and industry. From cutting-edge developments in hydrogen production, including electrolysis, photoelectrochemical processes, and biological systems, to sophisticated storage solutions like solid-state materials, liquid organic hydrogen carriers, and advanced composites, the chapter has highlighted hydrogen's pivotal role in the clean energy transition.

It also examined hydrogen's expanding footprint in utilization sectors, showcasing its impact in fuel cells and combustion processes and as a transformative agent in chemical synthesis. The narrative further extended into hydrogen's emerging roles, illustrating its potential in transportation, grid balancing, energy storage, and industrial applications, underscoring its versatility and adaptability across various sectors. The strategic importance of hydrogen technologies in achieving energy sustainability and climate goals cannot be overstated. As a clean, flexible, and efficient energy carrier, hydrogen stands at the forefront of the global shift toward a sustainable, low-carbon future. It bridges the gap between renewable energy sources and their integration into our daily lives, offering solutions for storage, transport, and decarbonization challenges that are critical for the energy transition.

This chapter's insights reflect a vision where hydrogen technologies are integral to our energy ecosystem, driving innovation, bolstering economic resilience, and paving the way for a sustainable environmental legacy. The collective advancements in hydrogen technology encapsulate a future where energy systems are interconnected, efficiency is maximized, and carbon emissions are significantly reduced, heralding a new era of energy that is clean, adaptable, and sustainable.

Conflict of interest

The authors declare no conflict of interest.

Author details

Carine Alves[1*], Gustavo Castro[2], Rodrigo Coelho[3] and Luciano Hocevar[1]

1 Federal University of Reconcavo of Bahia (UFRB), Feira de Santana, Brazil

2 Morais and Castro Engineering and Consultancy, Belo Horizonte, Brazil

3 SENAI CIMATEC, Salvador, Brazil

*Address all correspondence to: carine.alves@ufrb.edu.br

References

[1] Ding D, Wu X. Hydrogen fuel cell electric trains: Technologies, current status, and future. Applications in Energy and Combustion Science. 2024;**17**:100255. DOI: 10.1016/j.jaecs.2024.100255

[2] Sage V, Patel J, Hazewinkel P, Yasin QUA, Wang F, Yang Y, et al. Recent progress and techno-economic analysis of liquid organic hydrogen carriers for Australian renewable energy export—A critical review. International Journal of Hydrogen Energy. 2024;**56**:1419-1434. DOI: 10.1016/j.ijhydene.2023.12.269

[3] Zainal BS, Ker PJ, Mohamed H, Ong HC, Fattah IMR, Rahman SMA, et al. Recent advancement and assessment of green hydrogen production technologies. Renewable and Sustainable Energy Reviews. 2024;**189**:113941. DOI: 10.1016/j.rser.2023.113941

[4] Chelvam K, Hanafiah MM, Woon KS, Ali KA. A review on the environmental performance of various hydrogen production technologies: An approach towards hydrogen economy. Energy Reports. 2024;**11**:369-383. DOI: 10.1016/j.egyr.2023.11.060

[5] Dursun N. Production of biological hydrogen from Quinoa residue using dark fermentation and estimation of its microbial diversity. Heliyon. 2024;**10**:e25018. DOI: 10.1016/j.heliyon.2024.e25018

[6] Chu C, Wu K, Luo B, Cao Q, Zhang H. Hydrogen storage by liquid organic hydrogen carriers: Catalyst, renewable carrier, and technology—A review. Carbon Resources Conversion. 2023;**6**:334-351. DOI: 10.1016/j.crcon.2023.03.007

[7] Ahmed M, Dincer I. A review on photoelectrochemical hydrogen production systems: Challenges and future directions. International Journal of Hydrogen Energy. 2019;**44**:2474-2507. DOI: 10.1016/j.ijhydene.2018.12.037

[8] Rengui L. Latest progress in hydrogen production from solar water splitting via photocatalysis, photoelectrochemical, and photovoltaic-photoelectrochemical solutions. Chinese Journal of Catalysis. 2017;**38**:5-12. DOI: 10.1016/S1872-2067(16)62552-4

[9] Grahame A, Agyey-Zinsou K-F. Properties and applications of metal (M) dodecahydro-closo-dodecaborates (Mn=1,2B12H12) and their implications for reversible hydrogen storage in the borohydrides. Inorganics. 2018;**6**:106. DOI: 10.3390/inorganics6040106

[10] Sood S, Prakash O, Boukerdja M, Dieulot J-Y, Ould-Bouamama B, Bressel M, et al. Generic dynamical model of PEM electrolyser under intermittent sources. Energies. 2020;**13**:6556. DOI: 10.3390/en13246556

[11] Yodwong B, Guilbert D, Phattanasak M, Kaewmanee W, Hinaje M, Vitale G. AC-DC converters for electrolyzer applications: State of the art and future challenges. Electronics. 2020;**9**:912. DOI: 10.3390/electronics9060912

[12] O'Brien JE, Hartvigsen JL, Boardman RD, Hartvigsen JJ, Larsen D, Elangovan S. A 25kW high temperature electrolysis facility for flexible hydrogen production and system integration studies. International Journal of Hydrogen Energy. 2020;**45**:15796-15804. DOI: 10.1016/j.ijhydene.2020.04.074

[13] Bollman J, Pitchaimuthu S, Kuhnel MF. Challenges of industrial-scale testing infrastructure for green hydrogen technologies. Energies. 2023;**16**:3604. DOI: 10.3390/en16083604

[14] Bargash H, AlRashdi Z, Okedu KE, Desmond P. Life-cycle assessment study for bio-hydrogen gas production from sewage treatment plants using solar PVs. Energies. 2022;**15**:8056. DOI: 10.3390/en15218056

[15] Pal H, Ting J-M, Agarwal S, Ichikawa T, Jain A. The catalytic role of D-block elements and their compounds for improving sorption kinetics of hydride materials: A review. Reactions. 2021;**2**:333-364. DOI: 10.3390/reactions2030022

[16] Zacharia R, Rather SU. Review of solid state hydrogen storage methods adopting different kinds of novel materials. Journal of Nanomaterials. 2015;**2015**:1-18. DOI: 10.1155/2015/914845

[17] Niermann M, Drunert S, Kaltschmitt M, Bonhoff K. Liquid organic hydrogen carriers (LOHCs)—Techno-economic analysis of LOHCs in a defined process chain. Energy and Environmental Science. 2019;**12**:290-307. DOI: 10.1039/c8ee02700e

[18] Nachtane M, Tarfaoui M, Abichou MA, Vetcher A, Rouway M, Aâmir A, et al. An overview of the recent advances in composite materials and artificial intelligence for hydrogen storage vessels design. Journal of Composites Science. 2023;**7**:119. DOI: 10.3390/jcs7030119

[19] Laaouidi H, Tarfaoui M, Nachtane M, Trihi M. Energy absorption characteristics in hybrid composite materials for marine applications under impact loading: Case of tidal current turbine. International Journal of Energy Research. 2021;**45**:5894-5911. DOI: 10.1002/er.6210

[20] Tarfaoui M, Sassi S, Lagdani O, Nachtane M. Strain rate effects on the thermomechanical behavior of glass/polyester composite joints. Polymer Composites. 2021;**43**:36-51. DOI: 10.1002/pc_26355

[21] Alternative Fuels Data Center: How Do Fuel Cell Electric Vehicles Work Using Hydrogen? Available from: https://afdc.energy.gov/vehicles/how-do-fuel-cell-electric-cars-work [Accessed: April 1, 2024]

[22] De Bellis V, Piras M, Bozza F, Malfi E, Novella R, Gomez-Soriano J. Development and validation of a phenomenological model for hydrogen fueled PFI internal combustion engines considering Thermo-diffusive effects on flame speed propagation. Energy Conversion and Management. 2024;**308**:118395. DOI: 10.1016/j.enconman.2024.118395

[23] Carpenella S, Cecere D, Giacomazzi E, Quaranta I, Sorrentino G, Sabia P, et al. Large eddy simulation of hydrogen/air MILD combustion in a cyclonic burner. Applied Thermal Engineering. 2024;**244**:122733. DOI: 10.1016/j.applthermaleng.2024.122733

[24] Rambhujun N, Salman MS, Wang T, Pratthana C, Sapkota P, Costalin M, et al. Renewable hydrogen for the chemical industry. MRS Energy & Sustainability. 2020;**7**:1-16. DOI: 10.1557/mre.2020.33

[25] Kumar R, Singh R, Dutta S. Review and outlook of hydrogen production through catalytic processes. Energy and Fuels. 2024;**38**:2601-2629. DOI: 10.1021/acs.energyfuels.3c04026

[26] Ullah A, Hashim NA, Rabuni MF, Junaidi MUM. A review on methanol as a clean energy carrier: Roles of zeolite in improving production efficiency. Energies. 2023;**16**:1482. DOI: 10.3390/en16031482

[27] Zuben TWV, Moreira DEB, Germscheidt RL, Yoshimura RG, Dorretto DS, De Araujo ABS, et al. Is hydrogen indispensable for sustainable world? A review of H_2 applications and perspectives for the next years. Journal of the Brazilian Chemical Society. 2022;**33**:824-843. DOI: 10.21577/0103-5053.20220026

Section 2

Hydrogen Generation

Chapter 2

Black TiO_2 Material for Highly Efficient Green Hydrogen Production Enabled by Advanced Surface Engineering

Xiaodan Wang, Beibei Wang and Hao Shen

Abstract

Black TiO_2 (H-TiO_2), as a promising photoanode material, can be used for direct green hydrogen production without emissions to pollute the environment, but the reported surface engineering approaches for the preparation of black TiO_2 suffer from high temperatures, long processing time, or chemical residues, limiting its practical application in green hydrogen production. Here, we developed two advanced surface engineering technologies, overcoming the above limitations, to prepare a black TiO_2 photoanode that achieved the maximum photocurrent density reported to date. Moreover, we theoretically and experimentally revealed the formation mechanism of black TiO_2 and its enhanced photoelectrochemical (PEC) performance. These surface engineering technologies are not only suitable for the preparation of efficient photoanode materials for PEC hydrogen production but also play a beneficial and promoting role in the research and development of new materials for hydrogen fuel cells and hydrogen storage.

Keywords: photoelectrochemical green hydrogen production, black TiO_2, surface engineering technologies, formation mechanism, hydrogen fuel cells, hydrogen storage

1. Introduction

Photoelectrochemical (PEC) technology for solar fuel generation has garnered international recognition as a promising solution for addressing both energy and environmental challenges [1, 2]. This technology has the capability to mimic photosynthesis by converting solar energy into hydrogen energy and also has the potential to convert CO_2 into hydrocarbon fuel, making it an ideal solution for a sunshine economy.

The technology of PEC hydrogen production was initiated in 1972 by Fujishima and Honda, which discovered that single crystal TiO_2 can generate sufficient photovoltage when exposed to light to facilitate the splitting of water into hydrogen and oxygen [3]. This breakthrough demonstrated the potential of utilizing solar energy to directly produce green hydrogen without CO_2 emissions, which is regarded as the most promising method for achieving zero-carbon hydrogen in the future. Subsequently, semiconductor photocatalysts have triggered significant interest within the academic community.

TiO_2 has emerged as a prominent semiconductor photocatalyst among various materials due to its stable chemical properties, robust oxidation-reduction capabilities, and cost-effectiveness. However, as a wide bandgap semiconductor, TiO_2 can only absorb ultraviolet light within the solar spectrum. Enhancing the responsiveness of TiO_2 to visible light has become a key focus of research in this field. Recent studies have explored surface engineering technologies via doping of various elements to improve visible light absorption of TiO_2, yet the resulting photocurrent remains below 1 mA cm^{-2} [4]. The field of surface engineering for TiO_2 is currently grappling with the obstacle of achieving high solar-to-hydrogen conversion efficiency.

The global interest in the synthesis of black TiO_2 through high-pressure hydrogenation of TiO_2 has been significant. In 2011, Chen et al. presented an innovative approach to creating black TiO_2 in Science [5]. **Figure 1** visually represents the unique core-shell structure of black TiO_2 and an accompanying energy diagram illustrating a decreased band gap. The researchers verified that black TiO_2 displays a wider band for light absorption, allowing it to capture visible, ultraviolet, and infrared light, exhibiting outstanding photocatalytic characteristics. Since its inception, black TiO_2 has attracted considerable interest in the scientific community due to its unique combination of high sunlight absorption and exceptionally high photocatalytic abilities. However, its applications are hampered by impractical hydrogenation conditions (20 bar, 200°C, and 5 days). Developing new hydrogenation technologies at mild conditions is highly demanded.

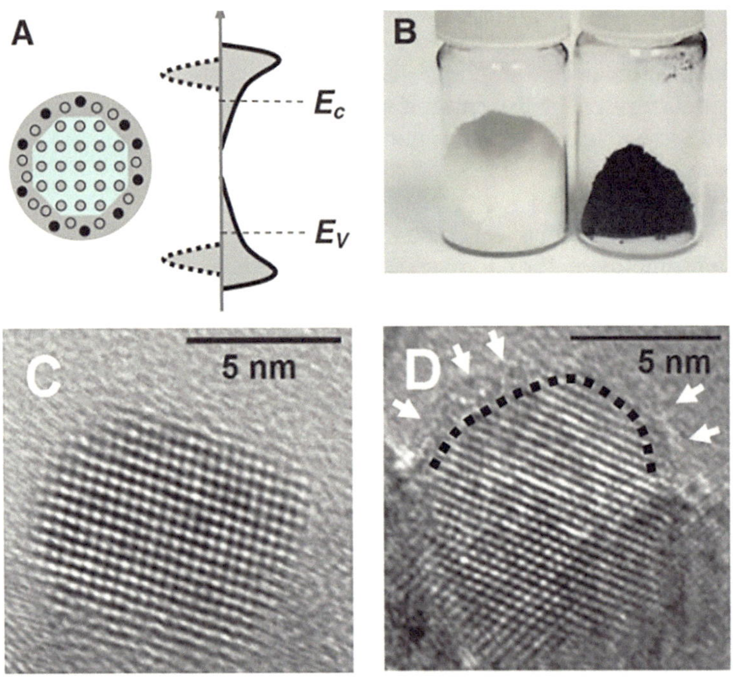

Figure 1.
(a) Core-shell structure and narrowing band gap models; (b) white pristine TiO_2 to black TiO_2; (c) and (d) high-resolution transmission electron microscope (HR-TEM) images of pristine TiO_2 and black TiO_2, respectively.

2. Recent advances in hydrogen technologies promoted by emerging hydrogenation methods

Hydrogenation technology is known as one of the most important material surface engineering technologies today, and here, we highlighted the latest research progress on how hydrogenation technologies can solve the core scientific problems of hydrogen-related technologies like PEC hydrogen production, hydrogen fuel cells and hydrogen storage.

2.1 PEC green hydrogen production

Wang et al. utilized high-temperature H_2 hydrogenation to prepare H-TiO_2 nanowires with significantly improved PEC performance for solar water splitting [6]. The H-TiO_2 nanorods treated at 350°C achieve the best photocurrent density of 2.4 mA cm^{-2} at 1.23 V_{RHE} compared to untreated nanorods, indicating efficient charge separation and transportation (**Figure 2a**). The incident photon to current efficiency (IPCE) analyses confirm that the enhancement of photocurrent density of H-TiO_2 is attributed to the enhanced photoactivity in the ultraviolet region (100%) and a significant increase in the donor density (3×), achieved through the creation of a high density of oxygen vacancies that act as electron donors. However, the harsh conditions of the high-temperature H_2 approach (400°C, 1 h) limit its practical application in PEC water splitting.

Xu et al. explored a Pd-catalyzed H hydrogenation strategy for reducing TiO_2, leading to the generation of Ti^{3+} species in TiO_2 and improved photocatalytic activity (**Figure 2b**) [7]. The hydrogenation process creates intrinsic defects in the materials, narrowing the band gap and enhancing solar energy utilization. The unique crystalline core/disordered shell structure of H-TiO_2 formed during the process contributes to improved photocatalytic activity and long-term stability. However, the high cost of noble metal and chemical residues in this hydrogenation method hinder its practical application in PEC hydrogen production.

2.2 Hydrogen fuel cells

The hydrogen fuel cell is divided into two chambers, which pass into H_2 and O_2, respectively; in the left chamber, under the catalytic Pt, H_2 dissociates to protons and

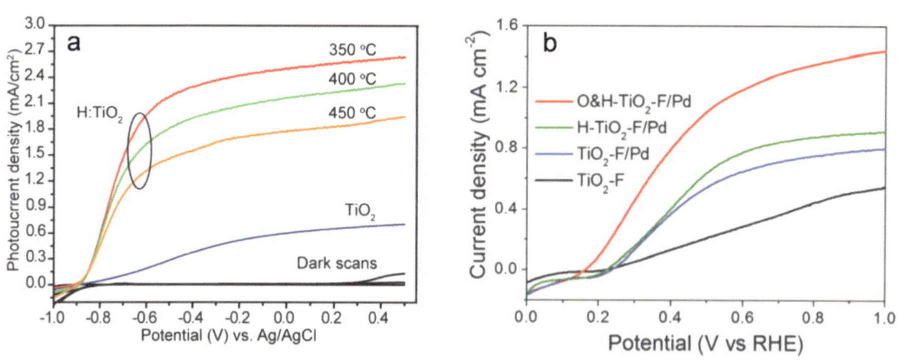

Figure 2.
Photocurrent density versus potential (J-V) measurements of H-TiO_2 via (a) high-temperature H_2 and (b) Pd-catalyzed H hydrogenations.

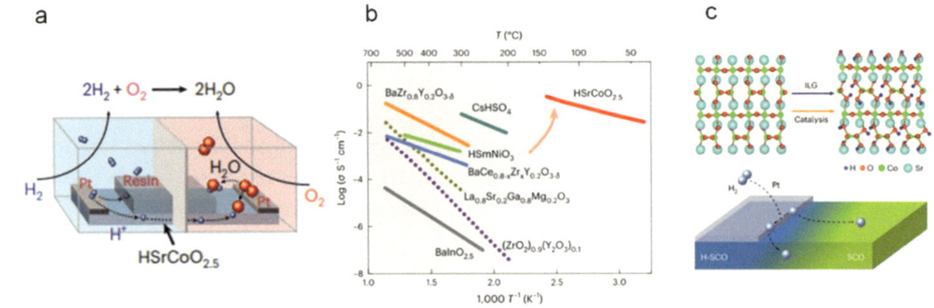

Figure 3.
(a) Scheme of hydrogen fuel cells. (b) Temperature-dependent proton conductivity: $HSrCoO_{2.5}$ vs. state-of-the art. (c) Scheme of noble metal catalyzed H hydrogenation and resulting organized oxygen vacancy channels in $HSrCoO_{2.5}$.

electrons ($2H_2 \rightarrow 4H^+ + 4e^-$), protons reach the counter electrode of the right chamber through the solid electrolyte, and at the same time, the electrons pass through the peripheral circuit to the counter electrode, so as to cooperate with the reduction reaction of protons to O_2 to generate H_2O ($O_2 + 4H^+ + 4e^- \rightarrow 2H_2O$), which completes the transfer of chemical energy to electrical energy (**Figure 3a**). It is evident that conversion efficiency depends strongly on the proton conductivity of solid electrolytes.

Solid oxide ionic conductors are utilized to serve as solid electrolytes in hydrogen fuel cells. Meanwhile, traditional low ionic conductors based on metal oxides typically necessitate temperatures exceeding 500°C to facilitate ionic transport (**Figure 3b**). Very recently, Lu et al. reported a new solid oxide proton conductor, $HSrCoO_{2.5}$, prepared by noble metal catalyzed H hydrogenation, exhibiting remarkably high proton conductivity within the temperature range of 40–140°C [8]. The proton conductivity in the specified temperature range ranged from 0.028 to 0.33 $S\,cm^{-1}$ (**Figure 3b**), which is ascribed to the elevated proton concentration and the organized oxygen vacancy channels facilitated by the noble metal catalyzed H hydrogenation (**Figure 3c**). However, the prohibitive expense of noble metals and the presence of chemical residues in the hydrogenation process pose significant obstacles to its feasibility for use in practical hydrogen fuel cells.

2.3 Hydrogen storage

Solid-state hydrogen storage materials, facilitating the storage and controlled release of hydrogen, have been a focal point of extensive research for numerous decades, establishing ambitious goals for their advancement in recent years. Developing such materials that can store an appreciable amount of hydrogen reversibly without having to use high pressure and/or low temperature is a significant challenge.

Izumi et al. prepared crystalline samples of the material $La_{1-x}Sr_xH_{3-x-2y}O_y$ ($0.1 \leq x \leq 0.6$, $y < 0.171$) through ball-milling, followed by annealing under high-pressure H_2 hydrogenation [9]. They conducted an examination of the samples at ambient temperature, determining their ability to facilitate the conduction of hydride ions at a notable rate. Subsequently, they assessed the efficacy of the material in an all-solid-state cell composed of Ti| $La_{1-x}Sr_xH_{3-x-2y}O_y$|$LaH_{3-\delta}$, manipulating the quantities of strontium and oxygen within the composition. Upon identifying an optimal strontium content of at least 0.2, they observed the complete conversion of titanium to titanium hydride (TiH_2) at a rate of 100%, indicating minimal wastage of hydride ions. Based on the results, they focus on enhancing performance and developing electrode materials with the capability to reversibly absorb and release hydrogen. This advancement is crucial for enabling the recharging of batteries and facilitating the storage and controlled release of hydrogen, thereby meeting the demands of hydrogen-based energy applications. While the above findings have made important advances in hydrogen storage, the high-temperature and high-pressure hydrogen hydrogenation method (400°C, 0.5 MPa, 12 h) they employed will prevent the development of its practical applications.

3. Advanced hydrogenation technologies for PEC green hydrogen production

The above-hydrogenated surface engineering technologies have shown great potential in PEC green hydrogen production, fuel cells, and hydrogen storage applications; however, the efficacy of their applications is hindered by impractical hydrogenation conditions and ambiguous hydrogenation mechanisms. Here, two advanced methods, *low-temperature H* [10] *and room-temperature H^+ hydrogenations* [11], are developed to address the challenging conditions associated with traditional high-temperature H_2 reductions and the constraints posed by chemical residues resulting from chemical reductions.

3.1 Low-temperature H hydrogenation

3.1.1 Experimental set-up

The concept of low-temperature hot wire H hydrogenation draws inspiration from previous research. Langmuir's study in 1912 investigated the efficiency of a tungsten filament in facilitating the dissociation of H_2 molecules into H atoms [12]. Experimental results demonstrated that when a tungsten filament is heated to temperatures exceeding 1000°C in a hydrogen-rich environment, H_2 molecules can be catalytically dissociated upon contact with the hot tungsten filament. In contrast to the H^+ hydrogenation approach, this method exclusively involves only hydrogen atoms rather than hydrogen ions.

Figure 4a depicts the experimental configuration utilized for the low-temperature H hydrogenation. Within a vacuum chamber, a parallel array of 10× tungsten wires, each measuring 600 mm in length, 0.53 mm in diameter, and 50 mm in space, was positioned, resulting in a total activated area of 600 × 450 mm^2. The samples were subjected to treatment at varying wire temperatures (1600, 1700, and 1800°C), with a separation of 75 mm between the wires and the samples.

The regulation of the flux rate of atomic hydrogen is contingent upon the wire temperature and pressure, with the pressure typically preset at 1 Pa. **Figure 4b** shows the plot of the flux of atomic hydrogen vs. wire temperature. It is evident that the flux of atomic hydrogen increases exponentially with increased wire temperature. This means that the low-temperature H hydrogenation [10] can achieve precise H-doping by tuning the wire temperature, demonstrating the superior features over the uncontrollable traditional high-temperature or high-pressure H_2 hydrogenations.

3.1.2 Hydrogenation-PEC performance relation

In this section, we examine in detail the effects of wire temperature on the structural, optical, electrical, and PEC properties of black TiO_2.

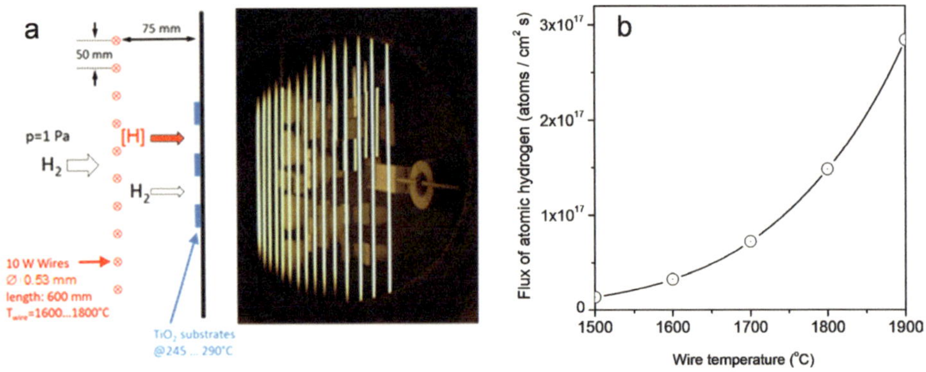

Figure 4.
(a) Experimental set-up of low-temperature H hydrogenation and (b) plot of the flux of atomic hydrogen vs. wire temperature.

3.1.2.1 Structure

Scanning electron microscopy (SEM) and HR-TEM were used to investigate the microstructure of H-TiO_2 nanorods treated at different wire temperatures. H-TiO_2 treated at T_{wire} = 1700°C closely resembles pristine TiO_2 in morphology. HR-TEM images show that pristine TiO_2 has a single crystalline structure (**Figure 5a–c**), while H-TiO_2 at T_{wire} = 1700°C has a core-shell structure with a 2 nm thick disordered shell (**Figure 5d–f**). Electron energy loss spectroscopy (EELS) analysis showed that the [O]/[Ti] ratio in pristine TiO_2 remains constant at about 2 throughout the nanorod, while in H-TiO_2 at T_{wire} = 1700°C, the ratio decreases gradually in the disordered region.

3.1.2.2 Optical properties

The optical absorption spectra of pristine TiO_2 and H-TiO_2 were analyzed to study the improved optical absorption properties of H-TiO_2. Results show a red shift in the band edge of H-TiO_2 with increasing wire temperatures and increased absorption in the visible spectrum (**Figure 6a**). Tauc plots indicate a band edge of 2.9 eV for H-TiO_2 compared to 3.0 eV for pristine TiO_2 (**Figure 6b**).

3.1.2.3 PEC properties

J-V measurements were conducted to investigate the improved PEC properties of H-TiO_2 nanorods (**Figure 7a**). The comparison of J-V curves of pristine TiO_2 and H-TiO_2 samples showed an increase in photocurrent density after treatment at T_{wire} = 1600°C, peaking at T_{wire} = 1700°C, and decreasing at T_{wire} = 1800°C. The effect of wire temperature on the conductivity of H-TiO_2 was studied using Mott-Schottky measurements (**Figure 7b**). Higher wire temperatures led to a decrease in the slope of H-TiO_2, indicating an increase in donor density and conductivity. The donor densities of H-TiO_2 were found to be 3× higher than those of pristine TiO_2, showing a trend of increased conductivity in H-TiO_2 samples with higher temperatures.

3.1.2.4 Ti^{3+} identification

X-ray photoelectron spectra (XPS) analysis of H-TiO_2 samples at different wire temperatures showed that the photocurrent density was highest at T_{wire} = 1700°C.

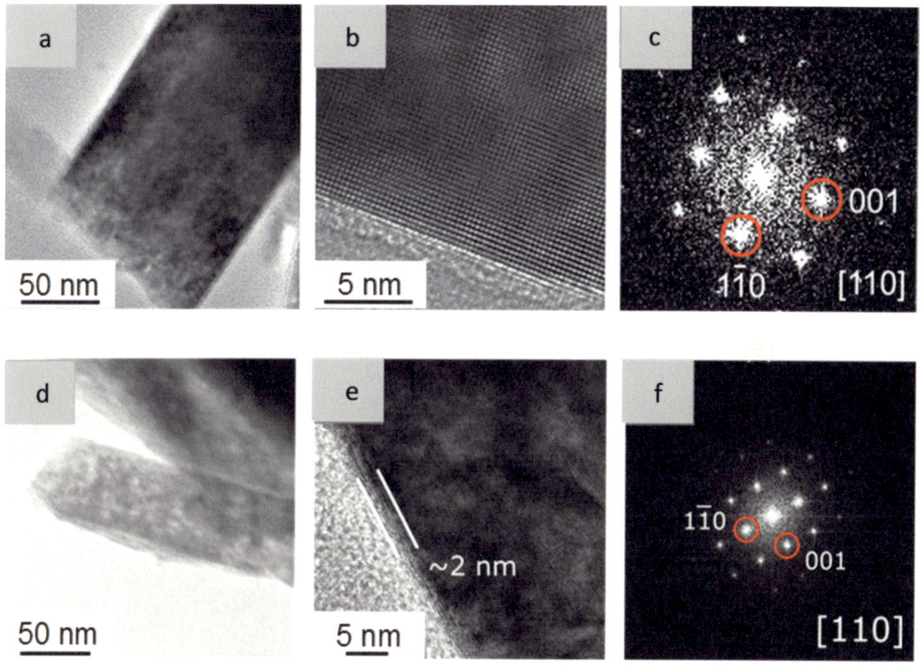

Figure 5.
TEM, HR-TEM images, and fast Fourier transform (FFT) patterns of (a–c) pristine TiO_2 and (d–f) H-TiO_2 nanorods treated at T_{wire} = 1700°C.

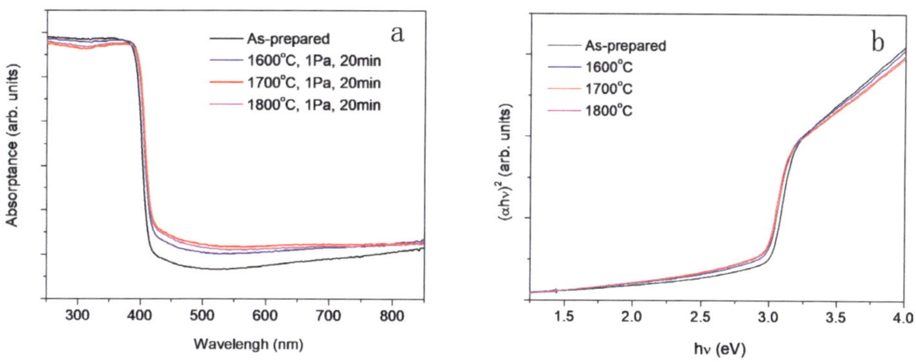

Figure 6.
(a) Optical absorption and (b) Tauc plots for pristine TiO_2 and H-TiO_2.

The O1s spectra revealed peaks at 530 and 531.5 eV, corresponding to TiO_2 and Ti-OH species (**Figure 7c**). The intensity of the Ti−OH peak increased with increasing wire temperature up to 1700°C but decreased at 1800°C. The reduced Ti-OH peak in H-TiO_2 at 1800°C suggests less Ti^{3+} presence. XPS Ti2p spectra for both samples show similar characteristics, with binding energies of 458.5 eV for $Ti2p_{3/2}$ and 464.3 eV for $Ti2p_{1/2}$, typical of TiO_2 (**Figure 7d**). A Ti-H peak at 456.5 eV, rather than metallic Ti at 453.8 eV, is observed in the sample treated at T_{wire} = 1800oC. The formation of Ti–H bonds in H-TiO_2 at 1800°C reduces Ti−OH bonds and decreases Ti^{3+} concentration, confirming the insight into improved photocurrent density of H-TiO_2 at T_{wire} = 1700°C.

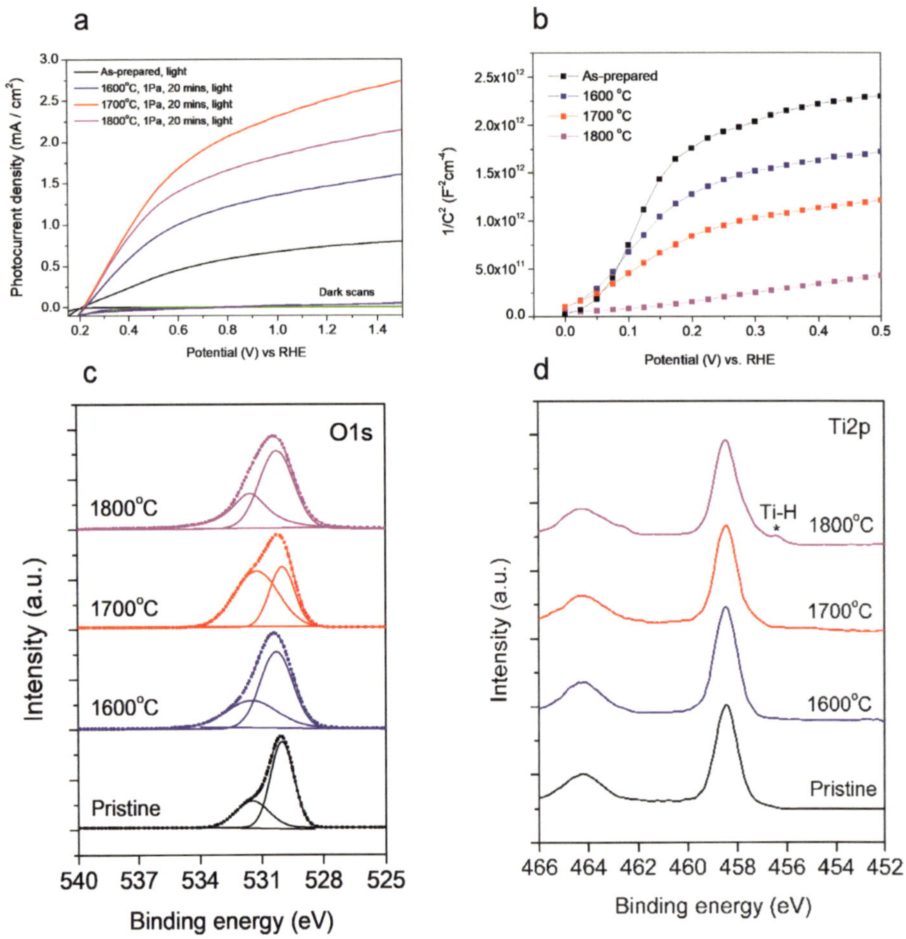

Figure 7.
(a) J-V curves of pristine TiO_2 and H-TiO_2 nanorods in dark and under solar illumination. (b) Mott-Schottky plots of pristine TiO_2 and H-TiO_2 nanorods. (c) O1s and (d) Ti2p XPS spectra of pristine and H-TiO_2 samples.

The results suggest that the low-temperature H hydrogenation process effectively facilitates the formation of a disordered shell on the H-TiO_2 nanorods, resulting in a decrease in band gap (3.0 → 2.9 eV), an increase in donor density (3×), and Ti^{3+} concentration, leading to the highest photocurrent density of 2.5 mA cm^{-2} at 1.23 V_{RHE} of H-TiO_2 at T_{wire} = 1700°C reported to date.

3.1.3 Formation mechanism

The low-temperature H hydrogenation process is delineated in the subsequent three stages: (a) hydroxylation of the TiO_2(110) surface through hydrogen adsorption; (b) subsurface diffusion of hydrogen; (c) subsurface hydrogenation.

Density functional theory (DFT) simulations suggest that H treatment facilitates the occupation of O_{3c} surface sites by H, subsurface diffusion, and subsurface hydrogenation through the $O_{3c} \rightarrow O_{sub}$ low-energy-barrier pathway (0.87 eV), leading to the thermodynamically favorable formation of a disordered shell. Conversely, H_2 treatment typically results in only submonolayer degrees of hydroxylation, with the

	[H]	H_2
Hydroxylation	Almost no barrier for adsorption of atomic hydrogen → low temperature is sufficient H_2 desorption is suppressed due to high-energy barrier	High barrier for dissociative adsorption molecular hydrogen → high temperature and flux are needed. High surface coverages are not thermodynamically favorable
Subsurface diffusion	Lower barrier for subsurface diffusion pathway is $O_{3c} \rightarrow O_{sub}$ subsurface diffusion is faster than H_2 and H_2O desorption at low temperatures	High barrier for subsurface diffusion pathway is $O_{2c} \rightarrow O_{3c} \rightarrow O_{sub}$ subsurface diffusion is very slow
Subsurface H-doping	Incorporation of H in TiO_2 lattice is thermodynamically favorable up to high concentrations	Incorporation of H in TiO_2 lattice is not thermodynamically favorable. At high temperatures, incorporation becomes even less favorable
Formation of disordered shell	Formation of disordered shell is thermodynamically favorable through $O_{3c} \rightarrow O_{sub} \rightarrow$ H-doping in TiO_2 lattice with relatively low barriers. *Formation pathway*: low temperatures are enough	$O_{2c} \rightarrow O_{3c} \rightarrow O_{sub} \rightarrow$ H-doping in TiO_2 is not thermodynamically favorable and kinetically suppressed. *Different formation pathway*: only H_2O desorption and formation of oxygen vacancies are thermodynamically favorable → high temperatures are needed

Table 1.
Different formation pathways of disordered shell: H vs. H_2.

occupation of O_{3c} sites being thermodynamically unfavorable. Consequently, subsurface diffusion from O_{2c} to O_{sub} is kinetically hindered during H_2 treatment; only H_2O desorption and the formation of oxygen vacancies are deemed thermodynamically feasible, necessitating high pressure and/or high temperatures for the process to occur (**Table 1**).

In conclusion, we have experimentally and theoretically confirmed that our low-temperature H hydrogenation method is indeed superior to the traditional H_2 hydrogenation method: the highest photocurrent density and highly efficient hydrogenation are achieved.

3.2 Room-temperature H^+ hydrogenation

3.2.1 Experimental set-up

Recent interest in H^+ hydrogenation using kinetic hydrogen ion species has increased because it does not require thermal activation. Traditional methods of H^+ hydrogenation are limited by the need for high temperatures or high power output from plasma equipment to generate high-energy hydrogen ions, leading to inefficient and uncontrollable processes as well as limited photocurrent density of <1 mA cm^{-2} at 1.23 V_{RHE} [13].

The magnetic field from an electromagnetic coil affects hydrogen ions and free electrons differently. Electrons change direction, but ions are not affected due to their greater mass (1000×). The magnetic field causes electrons to orbit in a circular path, increasing the chance of ionization and creating more hydrogen ions.

Inspired by the physical picture above, unlike traditional plasma hydrogenation methods, we used a circular electromagnetic coil in a radio frequency (RF) plasma physical vapor deposition (PVD) system to create a magnetic field (**Figure 8**). This field increases hydrogen ion concentration and decreases RF power and self-bias voltage, allowing for controlled low-energy hydrogen ion production for room-temperature H^+ hydrogenation [11].

A programmable recipe controlled the RF power, magnetic field strength, hydrogen flow rate, processing pressure, and treatment duration for the H^+ hydrogenation process. This included using RF power of 20 W, a magnetic field strength of 30 mT, a hydrogen flow rate of 50 sccm, and a processing pressure of 1.5 Pa to achieve a low self-bias voltage of −250V at the substrate electrode.

3.2.2 Hydrogenation-PEC performance relation

H-TiO_2 nanorods were synthesized at different time intervals (2.5, 5, 10, 20, and 40 mins) to investigate their structural, optical, electrical, and PEC properties.

3.2.2.1 Structure

SEM was used to study morphological changes in TiO_2 before and after H^+ treatment. No significant changes were observed in SEM images of pristine TiO_2 and H-TiO_2 treated at 5 and 40 mins. X-ray diffraction (XRD) patterns showed only a rutile phase with (101) and (002) peaks, indicating no phase change in either sample. HR-TEM images show that H-TiO_2 samples have a composite structure with a crystalline core and a disordered shell. The core remains single crystalline, with the shell thickness increasing from 2 to 5 nm as treatment time increases from 5 to 40 mins (**Figure 9**). The line-scan EELS data of [O]/[Ti] ratio changes in pristine TiO_2 and H-TiO_2 samples. Oxygen vacancies form in the shell of H-TiO_2 over time but not in the core of H-TiO_2 treated for 5 mins.

3.2.2.2 Optical properties

We recorded optical absorption spectra to compare the optical properties of pristine TiO_2 and H-TiO_2. The absorption band edges of H-TiO_2 shift toward red and

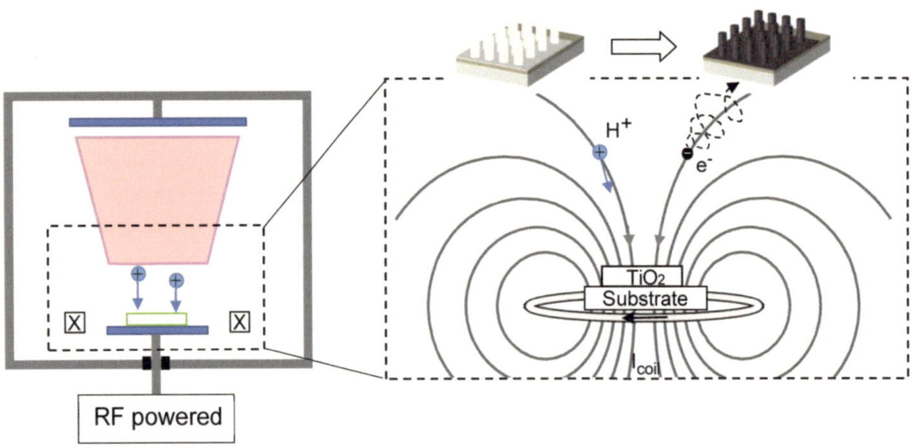

Figure 8.
Experimental set-up of room-temperature H^+ hydrogenation. The role of a circular electromagnetic coil in the H^+ hydrogenation process.

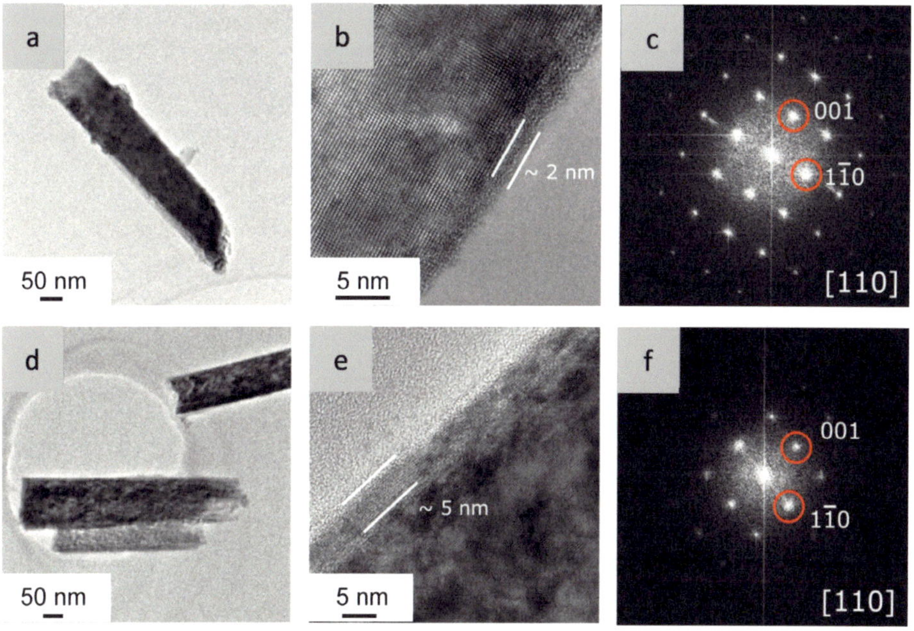

Figure 9.
TEM, HR-TEM, and FFT images of H-TiO_2 at (a–c) 5 and (d–f) 40 mins, respectively.

absorb more in the visible to infrared range with longer treatment time. The band gap decreases from 3.0 eV in pristine TiO_2 to 2.91 eV in H-TiO_2 after 5 mins of treatment.

3.2.2.3 PEC properties

Figure 10a shows that the photocurrent density of H-TiO_2 increases up to 5 mins of treatment time, then decreases. At 5 mins, a photocurrent density of 2.55 mA cm^{-2} at 1.23 V_{RHE} is achieved, the highest reported for H-TiO_2 studies. Increasing treatment time increases donor density and decreases depletion region width. The negative shift in flat band potential of H-TiO_2 at 2.5 and 5 mins is due to a higher donor density, shifting the Fermi level toward the conduction band. The largest band bending of 1.1 V is observed for H-TiO_2 at 5 mins, determined from M-S plots (**Figure 10b**).

3.2.2.4 Ti^{3+} identification

XPS analysis of the disordered shell showed the presence of Ti–O and Ti–OH in H-TiO_2. Ti^{3+} was detected in H-TiO_2 treated at 5 mins, with a strong reduction to Ti^{2+} observed in H-TiO_2 treated at 40 mins (**Figure 10c** and **d**).

Our experiment results confirmed that the highest photocurrent density of 2.55 mA cm^{-2} at 1.23 V_{RHE} of H-TiO_2 at 5 mins is due to a smaller band gap (2.91 eV), increased band bending (1.1 eV) and bulk defect suppression.

3.2.3 Formation mechanism

We studied how the penetration ratio of vertically impinging H^+ changes with initial kinetic energy $E_{kin,0}$. At the lowest $E_{kin,0}$ of 0.1 eV, all 15 H^+ were reflected, consistent with the above DFT study of low-temperature H hydrogenation predicting

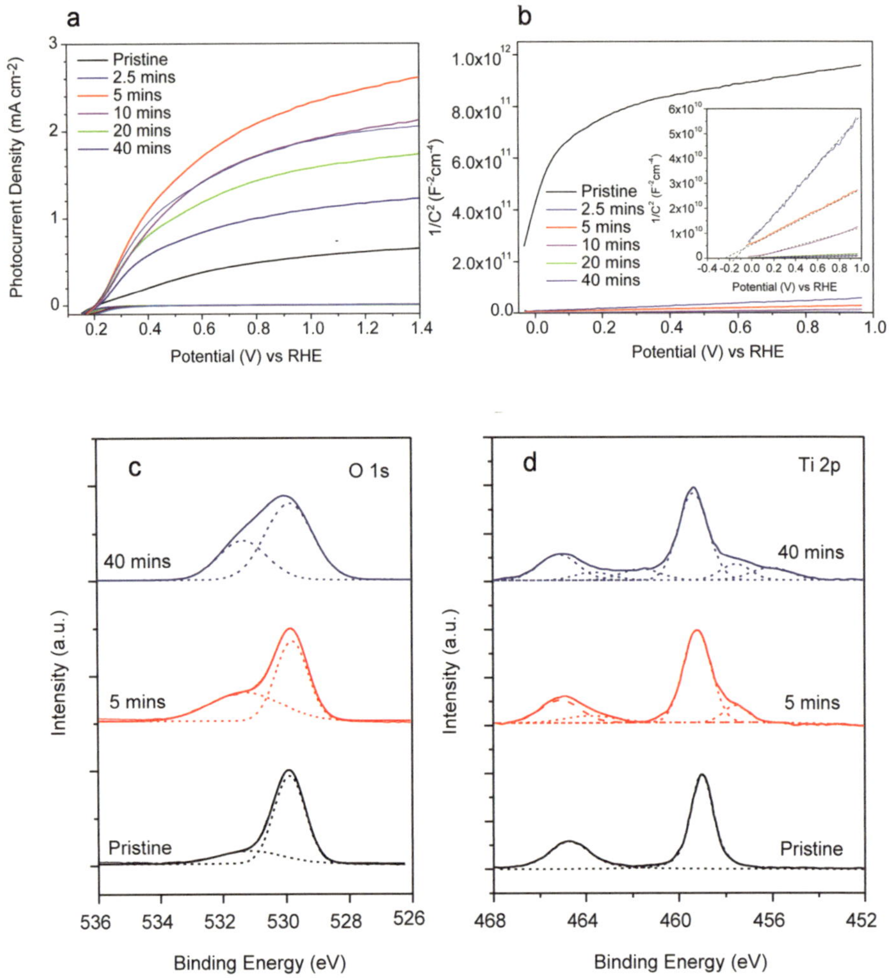

Figure 10.
(a) J-V curves of pristine TiO_2 and H-TiO_2 at different treatment times in dark and under solar illumination. (b) M-S plots of pristine TiO_2 and H-TiO_2 at different treatment times. (c) O1s and (d) Ti2p XPS spectra of pristine TiO_2 and H-TiO_2 at 5 and 40 mins, respectively.

a minimal energy barrier of 0.87 eV for subsurface diffusion. At $E_{kin,0}$ = 1 eV, 1 out of 15 H^+ bonded to a surface O ion. At $E_{kin,0}$ = 10 eV, 8 out of 15 H^+ penetrated the TiO_2 surface. At the maximum velocity of 80 eV, 10 out of 15 H^+ penetrated the TiO_2 surface, resulting in a 67% penetration ratio. H^+ with kinetic energies of at least 1 eV can penetrate the TiO_2 surface without needing thermal activation. The penetration ratio for H^+ with kinetic energies of at least 10 eV can reach approximately 0.5.

It is intriguing to compare how energetic H^+ particles interact with the TiO_2 surface versus thermal H species like H_2 and H in hydrogenation processes. Impinging thermal H_2 or H cannot directly penetrate; instead, hydrogen uptake in the subsurface region involves chemisorption and subsurface diffusion. Atomic H is more efficient than molecular H_2 for subsurface diffusion due to its ability to adsorb at higher energy surface sites, resulting in a lower energy barrier. However, efficient hydrogenation only occurs at high substrate temperatures. In our previous studies, we found that the lowest energy barrier for the movement of a chemisorbed H atom into the subsurface area is 0.87 eV and occurs near an O_{3c} site. In the H^+ case, as the initial kinetic energy increases, additional penetration pathways with higher energy barriers surrounding $O_{3c} \rightarrow O_{sub}$ are available. This phenomenon results in a rapid increase in the penetration probability with increasing initial kinetic energy, ultimately contributing to the high efficiency of room-temperature H^+ hydrogenation.

In summary, our study has experimentally and theoretically validated the superiority of our room-temperature H^+ hydrogenation over the thermal hydrogen hydrogenation approaches, as evidenced by the attainment of the highest photocurrent density and exceptional efficiency in hydrogenation processes.

4. Comparison of low-temperature H and room-temperature H^+ methods with state-of-the-art hydrogenations

Through a comparative analysis of the synthesis of black TiO_2 using the traditional hydrogenation techniques and the subsequent evaluation of photocurrent density, our two advanced hydrogenation methods, low-temperature H and room-temperature H^+, exhibit the highest reported photocurrent density (**Table 2**), while simultaneously effectively addressing the challenges associated with traditional H_2 hydrogenation, such as high temperature, high pressure, and prolonged treatment, as well as chemical residues from chemical reductions.

Table 3 shows that the hydrogenation conditions of the hydrogenation technologies recently applied to hydrogen fuel cells and hydrogen storage are not suitable for

Hydrogenation method	Temperature	RF power	Treatment time, pressure	Photocurrent density of H-TiO_2 at 1.23 V_{RHE}	Features and references
High-pressure H_2	200°C	N/A	5 days, 20 bar	N/A	x High pressure; prolonged treatment [5]
High-temperature H_2	400°C	N/A	1 h, 1 bar	2.4 mA cm^{-2}	x High temperature; prolonged treatment [6]
Low-temperature H	265°C	N/A	20 mins, 1 Pa	2.5 mA cm^{-2}	✓ Low temperature; short treatment [10]
Noble metal-catalyzed H	250°C	N/A	10 mins, 1 bar	1.5 mA cm^{-2}	x High cost: noble metals; chemical residues [7]
High-temperature H^+	425°C	200 W	1 h	0.9 mA cm^{-2}	x High temperature and power; prolonged treatment [13]
Room temperature H^+	25°C	20 W	5 mins, 1.5 Pa	2.55 mA cm^{-2}	✓ Room temperature; short treatment [11]

Table 2.
A comparison of the low-temperature H hydrogenation and room-temperature H^+ hydrogenation with state-of-the-art hydrogenations for PEC applications.

Hydrogenation method	Temperature	Treatment time, pressure	Applications	Features and references
Noble metal-catalyzed H	250°C	10 mins	Hydrogen fuel cells	x High cost: noble metals; chemical residues [8]
High-pressure H_2	400°C	12 h, 0.5 MPa	Hydrogen storage	x High pressure; prolonged treatment [9]

Table 3.
Reported hydrogenations for hydrogen fuel cells and storage applications.

practical applications, and it is evident that the two advanced hydrogenation technologies can compensate for their shortcomings.

5. Conclusions

In summary, we highlighted the advancements in surface engineering-hydrogenation technology for use in PEC hydrogen production, hydrogen fuel cells, and hydrogen storage. We presented the operational principles and experimental set-up of our invented hydrogenation technologies, and subsequently investigated the correlation between hydrogenation parameters and the structural, optical, electrical, and PEC properties of materials through combined experimental and theoretical studies. Finally, we conducted a comparative analysis of the benefits and drawbacks of the two hydrogenation technologies in relation to reported hydrogenation technologies. Our findings suggest that these two advanced technologies hold significant potential for widespread applications in various hydrogen-related technologies, such as PEC hydrogen production, hydrogen fuel cells, and hydrogen storage.

Acknowledgements

X. Wang acknowledges the support of the research fund by the Institute of Carbon Neutrality, ShanghaiTech University. H. Shen and B. Wang thank the support of the research fund by Yiwu Research Institute, Fudan University.

Conflict of interest

The authors declare no conflict of interest.

Author details

Xiaodan Wang[1], Beibei Wang[2] and Hao Shen[2]*

1 Institute of Carbon Neutrality, ShanghaiTech University, Pudong, Shanghai, China

2 Yiwu Research Institute, Fudan University, Yiwu, Zhejiang, China

*Address all correspondence to: shenhao@ywfudan.cn

References

[1] Yang W, Prabhakar RR, Tan J, Tilley SD, Moon J. Strategies for enhancing the photocurrent, photovoltage, and stability of photoelectrodes for photoelectrochemical water splitting. Chemical Society Reviews. 2019;**48**(19):4979-5015. DOI: 10.1039/C8CS00997J

[2] El Ouardi M, El Idrissi A, Arab M, Zbair M, Haspel H, Saadi M, et al. Review of photoelectrochemical water splitting: From quantitative approaches to effect of sacrificial agents, oxygen vacancies, thermal and magnetic field on (photo) electrolysis. International Journal of Hydrogen Energy. 2024;**51**:1044-1067. DOI: 10.1016/j.ijhydene.2023.09.111

[3] Fujishima A, Honda K. Electrochemical photolysis of water at a semiconductor electrode. Nature. 1972;**238**(5358):37-38. DOI: 10.1038/238037a0

[4] Zhang X, Zhang S, Cui X, Zhou W, Cao W, Cheng D, et al. Recent advances in TiO_2-based photoanodes for photoelectrochemical water splitting. Chemistry, an Asian Journal. 2022;**17**(20):e202200668. DOI: 10.1002/asia.202200668

[5] Chen XB, Liu L, Yu PY, Mao SS. Increasing solar absorption for photocatalysis with black hydrogenated titanium dioxide nanocrystals. Science. 2011;**331**(6018):746-750. DOI: 10.1126/science.1200448

[6] Wang GM, Wang HY, Ling YC, Tang YC, Yang XY, Fitzmorris RC, et al. Hydrogen-treated TiO_2 nanowire arrays for photoelectrochemical water splitting. Nano Letters. 2011;**11**(7):3026-3033. DOI: 10.1021/nl201766h

[7] Xu Y, Zhang C, Zhang L, Zhang X, Yao H, Shi J. Pd-catalyzed instant hydrogenation of TiO_2 with enhanced photocatalytic performance. Energy & Environmental Science. 2016;**9**(7):2410-2417. DOI: 10.1039/C6EE00830E

[8] Lu N, Zhang Z, Wang Y, Li H-B, Qiao S, Zhao B, et al. Enhanced low-temperature proton conductivity in hydrogen-intercalated brownmillerite oxide. Nature Energy. 2022;**7**(12):1208-1216. DOI: 10.1038/s41560-022-01166-8

[9] Izumi Y, Takeiri F, Okamoto K, Saito T, Kamiyama T, Kuwabara A, et al. Electropositive metal doping into lanthanum hydride for H−conducting solid electrolyte use at room temperature. Advanced Energy Materials. 2023;**13**(43):2301993. DOI: 10.1002/aenm.202301993

[10] Wang X, Mayrhofer L, Hoefer M, Estrade S, Lopez-Conesa L, Zhou H, et al. Facile and efficient atomic hydrogenation enabled black TiO_2 with enhanced photo-electrochemical activity via a favorably low-energy-barrier pathway. Advanced Energy Materials. 2019;**9**(33):1900725. DOI: 10.1002/aenm.201900725

[11] Wang X, Mayrhofer L, Keunecke M, Estrade S, Lopez-Conesa L, Moseler M, et al. Low-energy hydrogen ions enable efficient room-temperature and rapid plasma hydrogenation of TiO_2 nanorods for enhanced photoelectrochemical activity. Small. 2022;**18**(46):2204136. DOI: 10.1002/smll.202204136

[12] Langmuir I. A chemically active modification of hydrogen 5. Journal of the American Chemical Society. 1912;**34**(10):1310-1325. DOI: 10.1021/ja02211a004

[13] Tian Z, Cui H, Zhu G, Zhao W, Xu J, Shao F, et al. Hydrogen plasma reduced black TiO_2-B nanowires for enhanced photoelectrochemical water-splitting. Journal of Power Sources. 2016;**325**:697-705. DOI: 10.1016/j.jpowsour.2016.06.074

Section 3

Hydrogen Storage

Chapter 3

Hydrogen Storage Vessels of Type 4 and Type 5

Kheireddin Kadri, Abir Ben Abdallah and Sébastien Ballut

Abstract

This chapter explores the optimization of type 4 pressure vessels used for hydrogen storage, focusing on carbon fiber-reinforced composites produced through filament winding. Many studies delve into the intricacies of the winding process to enhance the structural integrity of the vessels. Progressive failure analysis is employed to identify potential weak points and failure modes, guiding the development of optimal designs for improved safety and performance. Additionally, the chapter highlights the importance of considering recycling strategies in the design phase to address environmental concerns associated with composite materials. The findings contribute to advancing sustainable practices in the production and life cycle management of hydrogen pressure vessels.

Keywords: composite storage vessels, filament winding, progressive failure, filament winding, recycling

1. Introduction

Composite materials including high-performance fibers (glass, carbon, aramide, or organic) have emerged as alternative materials. Those composites with a polymer matrix, and fibers such as carbon fiber are viable candidates known as composite fiber-reinforced polymers (CFRP). They are being used in making hydrogen storage vessels due to their lightweight, high-strength, and corrosion-resistant properties.

In this chapter, we will expose the main storage vessel's features, which are the key stones of a high-performance hydrogen storage vessel. We will build this chapter similarly to the industrial process in which a type 4–5 storage vessel is built. In the first section, we will expose why a storage vessel made of composite is predominant. Then, in the second section, we will explain the choice of epoxy resin as a major material for designing the tank. The third section will deal with the process of filament winding. Here, the process of continuous fiber reinforcement will be exposed. The winding filament methodology will be highlighted. The choice of carbon fiber for winding over all other types of fibers will be justified with regard to the mechanical performance, among many other factors. The last section will treat the unavoidable question of recycling. The end-of-life of each component of the tank will be reviewed.

2. Why the composite storage vessel: a game changer

The composite storage vessels (CSV) have gained importance since the moment that an important part of energy transition will be built around hydrogen storage [1]. In fact, it is highly appealing for autonomy in the car industry to store a large amount of gas by just applying high compression. The most attractive feature of hydrogen tanks is their lightweight.

2.1 Description and orders of magnitudes

The CSVs of type (4) see 1 are refer to references [1, 2] used to store gaseous hydrogen under pressure of 700 bar. Physical consideration concerning the Thermos-chemical properties like phase diagram and energy densities can be found in this paper [3]. Those CSVs are made of the following components:

1. Cylindrical composite shell: tasked with facilitating the mechanical structuring of the reservoir, the external wall consists of a shell measuring several centimeters in thickness, composed of a carbon fiber-based composite material (see **Figure 1**) and a polymer resin.

2. Liner: inner wall, with a thickness of a few millimeters, serves the purpose of containing hydrogen at 700 bar and 23 K and providing sealing; it bears multiple tens of thousands of refills without developing cracks. Thermoplastic liners are commonly employed to ensure gas tightness and chemical compatibility. Commonly utilized thermoplastic liners include polyethylene (PE), polypropylene (PP), high-density polyethylene (HDPE), and polytetrafluoroethylene (PTFE).

3. Boss: At each end of the liner are two metal fittings called boss, which serve to connect the tank to the fuel cell and filling system. Functioning to provide mechanical structuring of the tank.

4. Carbon filament: this part is the most crucial because it will decide for the future behavior of the entire CSV [4] (see next paragraph §5). The carbon fiber takes the form of thin filaments with diameters ranging from 5 to 10 μm made of over 90% carbon. It is derived through the carbonization process of fibers from a

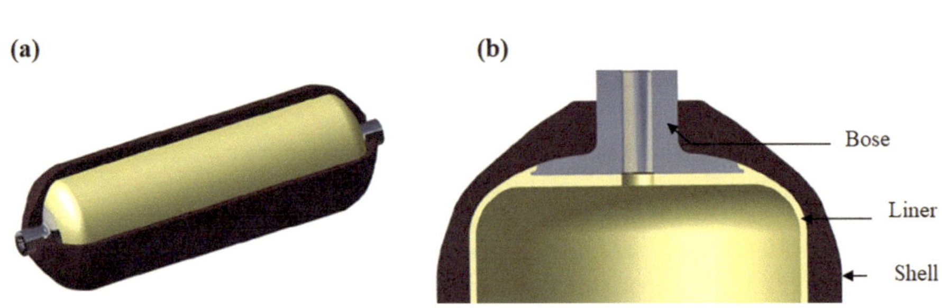

Figure 1.
(a) Perspective representation. (b) Vertical section representation of a type IV tank composed of metallic bases in gray to connect the tank to the fuel cell and filling system, in yellow a polymer material liner for hydrogen sealing, and in black of a composite shell with carbon fiber for mechanical structuring (adapted from Ref. [2]).

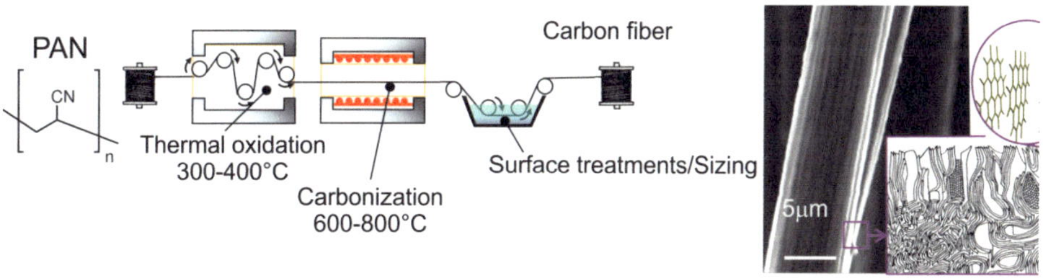

Figure 2.
Fabrication steps of carbon fibers from polyacrylonitrile, oxidation, and cross-linking of the PAN fibers, pyrolysis occurring at the higher temperatures under inert atmosphere or vacuum conditions (adapted from Ref. [5]).

CSVs	Pressure (bar)	Volume (L)	Weight (kg)	Volume density(MJ/L)	Cost ($/kg)
Type IV	700	62	55	6	600
Type V	700	62	50	8	800

Table 1.
Projected performance and cost of CSVs of both type IV and type V.

polymer known as a precursor, with polyacrylonitrile (PAN) being the most commonly utilized as a precursor (see next **Figure 2**).

Typically, for a hydrogen tank containing 62 L and 2.5 kg of hydrogen, the outer diameter measures approximately 40 cm, with a length of 90 cm; carbon fiber constitutes 60% of the total mass of the tank. Its total mass is approximately 55 kg, of which 33 kg are carbon fibers. One kilogram of hydrogen provides a range of 100 kilometers; therefore, each vehicle requires at least 5 kilograms of hydrogen to achieve a range of at least 500 kilometers. Knowing that at 700 bar and 15°C, a volume of 25 liters is required to store 1 kg of gaseous hydrogen, an internal storage volume of 125 L is needed. In the industry of automobile, it will be implemented in the form of two CSVs.

Table 1 summarizes the main features of hydrogen CSTv of type IV.

2.2 Physical justification: permeability

The physical basement of choosing composite is *permeability effect*. Two mechanisms contribute to gas leakage in composite materials: diffusion and microcracking, according to Humpenëoder et al. [6].

2.2.1 Diffusion

This effect is permanently present as long as both fiber and material have a certain rate of porosity. Many studies have been conducted to shed light on it. In the early 2000, experimental works [7] on polymer diffusion were done on three semi-crystalline polymers: polyethylene (PE), polyamide 11 (PA11), and poly(vinylidene fluoride) (PVF2). They were studied in the presence of helium (He), argon (Ar), nitrogen (N2), methane (CH4), and carbon dioxide (CO2) for temperatures ranging from 40° to 80°C in the case of PE, and from 70° to 130°C for both other. The applied pressures were, in the majority of tests, 10 MPa for He, Ar, N_2, which is far from the 70 MPa standard pressure of the present hydrogen CSVs. In the last years, many interesting experimental works have been oriented toward realistic condition of pressure [8, 9]. It has been found by Fujiwara et al. that the polymers used as liners (LDPE, LLDPE, HDPE, and MDPE) suffered fracture during the decompression process after hydrogen exposure was found. The permeability coefficient decreased with the decrease of diffusion coefficient under higher pressure condition. The second main result pointed out was that the shrinkage in free volume caused by hydrostatic effects of the applied hydrogen gas pressure decreases diffusion coefficient, resulting in the decrease of permeability coefficient with the pressure rise. In the following **Figure 3** is summed-up the main experimental result obtained by Fujiwara et al. [8].

They are bringing proofs that at steady-state high-pressure hydrogen gas permeation test (HPHP) under 90 MPa, the polypropylene (PP) suffers more destruction at higher exposure pressures, and the destruction in materials with smaller crystallinity was more severe. Another main fact is that hydrogen gas permeability increases with hydrogen pressure, but the increase ratio slows down with increasing pressure. Besides, the compressive effect of free volume after the application of hydrogen lowered the gas diffusion, and the permeation coefficient was also reduced in high-pressure environments.

In the last years, a noticeable study run by Conde-Wolter and co-workers [9] was achieved concerning permeability. **Figure 4** schematizes the experimental set-up that was used to measure permeability.

High-pressure hydrogen permeation tests were carried out on various thermoplastic matrix materials and on continuous fiber-reinforced thermoplastic composites (PA6, PA12, PA410, PPA, and PPS). Thin layers of liner made of ethylene vinyl alcohol (EVOH) thermoplastic films were added to PA6 composites. They studied how they affected the permeation rate. They also conducted pressure tests and examined micrographs to check for any manufacturing issues, pore size, and other defects like microcracks. Their experiments found that these thermoplastic composites can achieve low permeation rates as long as the microstructure remains intact.

The numerical approach is also very promising. According to this article [10], it is possible to determine effective diffusion coefficients numerically for CSVs. In

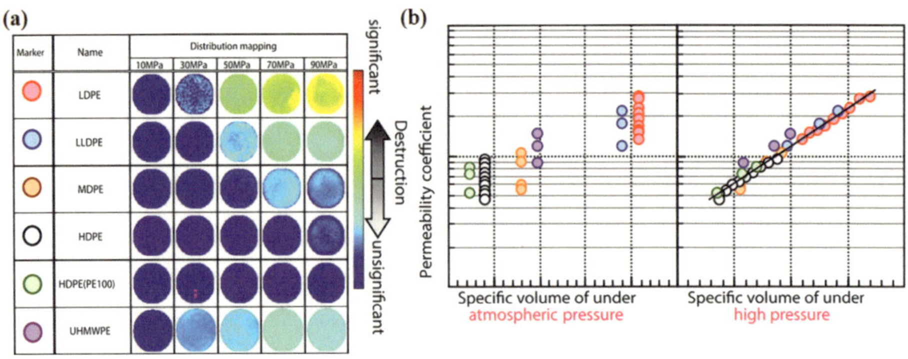

Figure 3.
(a) Matrix representation of n of the transmitted light image of six different polymers. Evaluation of destruction revealed when exposed to 10–90 MPa hydrogen for 24 h at 30°C. (b) Hydrogen gas permeation characteristics Relationship with specific volume in atmospheric pressure environment and in high-pressure environment (adapted from Ref. [8]).

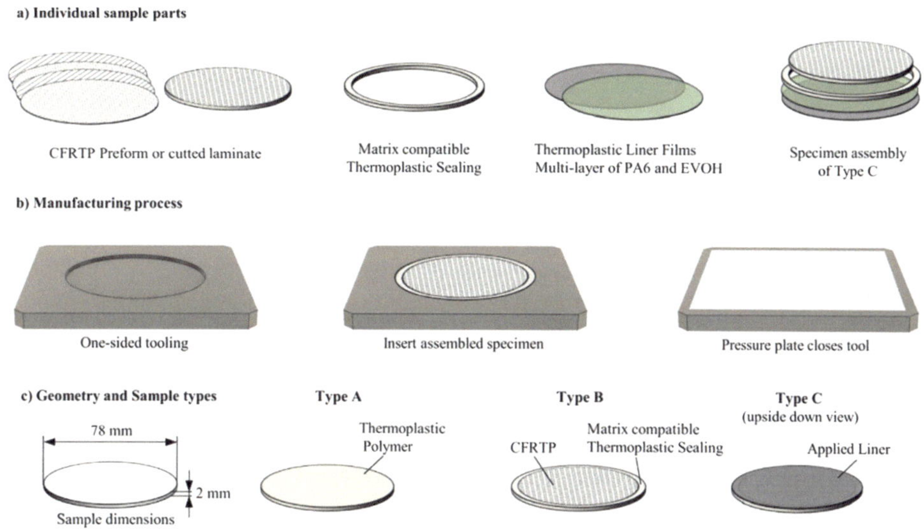

Figure 4.
(a) Material composition – thermoplastic liner films only necessary for type C specimens. (b) Manufacturing process. (c) Sample dimensions and different type of samples – type A – unreinforced samples of matrix polymer; type B – CFRTP sample with sealing, type C – CFRTP sample like type B with applied multilayer liner (adapted from Ref. [9]).

a framework of inhomogeneous composite laminate, three different homogenization methods were used. In order to assess the effective permeation coefficients for hydrogen permeation through composite laminates, the Wiener bounds, the Hashin-Shtrikman bounds, and a numerical finite element calculation of a representative volume element (RVE) were compared. It was possible to estimate the influence of the crack volume on the effective permeability. The results show that microcracking in a composite pressure vessel significantly affects the leak tightness, even if the crack volume is very small. The pressure vessel should therefore be free of matrix cracks if a liner-less design is aimed at. The crack volume fraction and the assumed permeability coefficient of the crack have no significant influence on the real permeability coefficient as long as they are within a plausible range. The main influencing factors are the fiber volume fraction and the crack configuration, respectively (**Figure 5**).

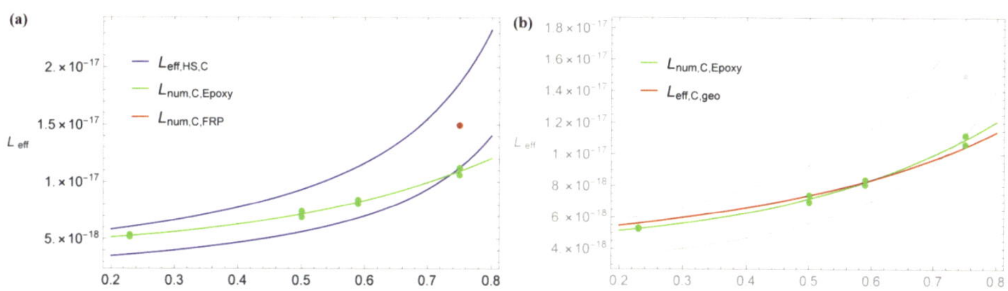

Figure 5.
(a) Influence of the crack configuration on the effective permeability. (b) Influence of the crack configuration on the effective permeability. Approximation of the numerical calculation is done using the geometric mean. In gray the bounds by use of parallel series and the numerically calculated result of the RVE in (a) (adapted from Ref. [10]).

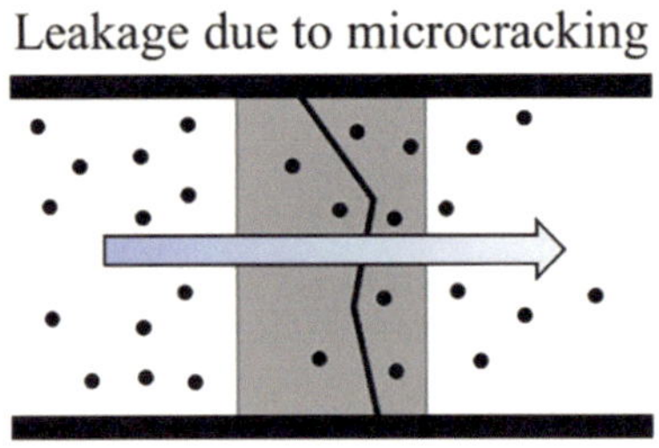

Figure 6.
Schematizing of the leakage mechanism due to intersecting matrix microcracks (adapted from Ref. [11]).

2.2.2 Microcracking

Figure 6 represents the enhancing effect of microcracking on hydrogen particle diffusion.

Basically, matrix microcracking occurs when mechanical and thermal stresses are applied to the composite laminate. These cracks form within each layer and run parallel to the fibers. Over time, these cracks can create pathways through the vessel's walls, allowing gas to escape. This is the initial sign of failure in a composite tank. While these cracks may not immediately cause the tank to fail completely, they weaken the material and can eventually lead to catastrophic failure. Therefore, it is crucial for designers to comprehend how these cracks form and their impact on the tank's performance.

This phenomenon has been found to result in significantly higher leakage compared to diffusion [12, 13]. Various factors, such as temperature, pressure, matrix properties, fiber distribution, fiber type, and ply stacking sequence, affect permeability. However, material-based properties only cause a minor change in permeability. Understanding permeation in polymers at high pressures, like 700 bar, remains limited. Testing at such pressures is challenging due to equipment limitations, and most measurements are conducted at lower pressures. Fujiwara et al. [8] addressed this by creating a 900 bar permeability cell, but it focused on polymers, not composites. Additionally, typical permeation tests are performed with the sample unloaded, which is not representative of the actual operating conditions of a pressurized vessel.

3. Matrix materials: epoxy resin

The majority of existing hydrogen pressure vessel (HPV) designs are manufactured using a thermoset matrix. Those classes of material are easier and more reliable to produce [14]. Due to their lightweight characteristics, resistance to corrosion, and gas permeability, resin epoxy is considered as the market standard thermoset. In this article [15], the authors focused on the specific application of epoxy resin in cryocompressed hydrogen storage vessels. The incorporation of polyethylene glycol-modified epoxy resin is highlighted for its role in enhancing the properties of the composite layer. This modification showcases the adaptability of epoxy resin to address the unique challenges associated with cryogenic conditions, underscoring its importance in ensuring the integrity and durability of hydrogen storage vessels. The study demonstrates how tailored modifications to epoxy resin contribute to the improvement of crucial properties, such as thermal stability and mechanical strength.

4. Carbon fiber synthesis

Carbon fiber is a material composed of thin filaments with diameters ranging from 5 to 10 μm, consisting primarily of carbon (over 90%). Carbon fiber is produced through the carbonization of precursor fibers, typically polyacrylonitrile (PAN), which accounts for over 90% of global production. In the next **Figure 2** a PAN is synthesized via radical polymerization of acrylonitrile and then spun through a spinneret with thousands to hundreds of thousands of holes. The resulting fibers undergo a series of thermal treatments to convert them into carbon fibers. Oxidation heating renders the fibers infusible, followed by an initial carbonization treatment under an inert atmosphere to obtain carbon fibers. For higher modulus fibers, a third thermal treatment called graphitization is performed at temperatures around 2000–2500 °C. Additionally, a thin polymer layer, known as sizing, is applied to the fiber surface to protect it during storage and transportation and to enhance the interface with the matrix during composite material fabrication.

Carbon fibers possess superior mechanical properties, including tensile strength and Young's modulus, compared to most materials, making them highly desirable for various applications. They are primarily used to manufacture composites with high mechanical properties by integrating them into a polymer matrix to bind and maintain the fibers. The quality and properties of these composites depend not only on the fiber and matrix quality and nature but also on the architecture used. Architecture refers to controlling the fiber orientation within the material to achieve desired mechanical properties in multiple directions. Since fibers act as one-dimensional elements, they only contribute their properties along their length, resulting in anisotropy in the composite. To obtain composites with desirable mechanical properties in multiple directions, it is essential to control fiber orientation within the material. This architecture is typically achieved through braiding, weaving, or winding techniques.

One crucial factor in fiber selection is the size effect, as investigated by Hwang et al. [16]. They found that as the component size increases, there is a reduction in the strength of the fiber, affecting vessel performance. They also developed the ring burst test, which accurately predicts failure strain results similar to full-scale vessels. Cohen et al. [17] demonstrated the impact of fiber volume fraction (FVF) on vessel strength, noting that increasing FVF in hoop plies improves ultimate strain-to-failure and failure pressure. According to this extensive review [18], there is a classification of the different mechanical properties of carbon fibers based on their chemical synthesis methods. **Table 2** summarizes the most important carbon fiber used for hydrogen storage vessels.

Carbon fiber	Tensile strength(MPa)	Young's modulus(GPa)	Elongation at break(%)	Ref
T700	4900	230	1.8	[19]
M40	3650	230	1.4	[20]
IM7	4500	275	1.5	[21]
T800	4600	240	1.6	[22]
T800S	4950	230	2.0	[23]

Table 2.
Mechanical properties of carbon fibers used in high-pressure storage vessels.

5. Filament winding for carbon fiber

This aspect of industrial processing was a major ax of development during the last decades [24]. It involves wrapping around a composite cylinder (named mandrel) in addition to the two hemispherical domes at each side (see **Figure** 7). This process, known since the 1940s, can be described in three major steps

1. Filament winding holds the fiber under tension during placement. It will only wrap around convex surfaces.

2. The fiber band is kept continuous throughout the winding process, which causes excess thickness build-up around the polar boss.

3. Placed under tension, the fibers self-align to geodesic paths (see next section). To achieve non-geodesic paths, friction is required to prevent fibers from slipping [24].

This continuous process of depositing tapes of carbon fibers will end up formatting multiple ply. Each of those stratified ply is different from the previous one by an angle value that will make a sequence.

Figure 8 illustrates a sequence of ply that make up the external reinforcement of CSV. Many approaches [11, 26] were utilized to investigate the effects of winding angle on filament-wound pressure vessels.

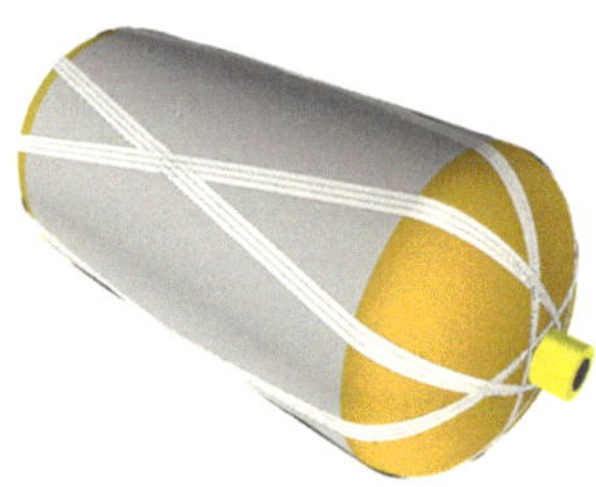

Figure 7.
Illustration of step of automated deposition of one ply of filament.

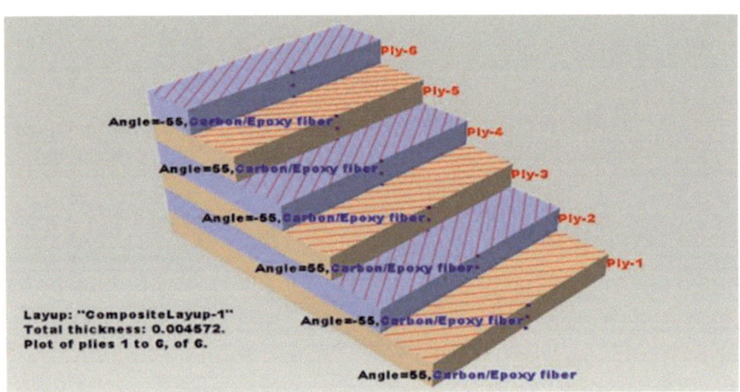

Figure 8.
The stacking sequences for 55° winding angle (adapted from Ref. [25]).

5.1 Winding patterns: laminated ply

A geometric approach for filament winding pattern generation on surface of revolution has been developed [27]. When winding fibers are applied on a cylindrical surface, there are three common types of paths: geodesic, non-geodesic, and semi-geodesic.

- Geodesic: These paths are the shortest routes between two points on the surface and are stable, requiring no external force to stay in place.

- Non-geodesic: Paths offer more design flexibility but are unstable and need frictional forces to prevent slipping. The choice between them depends on the shape of the CSV.

- Semi-geodesic: The way of winding involves a slight deviation from the geodesic path, which depends on the necessary friction to maintain the fiber in its intended position. This technique, also known as stable non-geodesic winding, offers flexibility in optimizing fiber paths.

Geodesic winding is preferred for most designs. Geodesic winding has zero lateral force on the fiber, while non-geodesic winding has some lateral force. Semi-geodesic winding is a stable variation of non-geodesic winding, offering flexibility while still maintaining stability.

Wet and dry winding are two methods used for winding fibers. In wet winding, fibers are soaked in resin and wrapped around a rotating mandrel, while in dry or prepreg winding, pre-impregnated fiber tows are used. Wet winding is commonly preferred, especially for making filament-wound composite cylinders, due to its advantages such as lower material costs, shorter winding periods, and easily

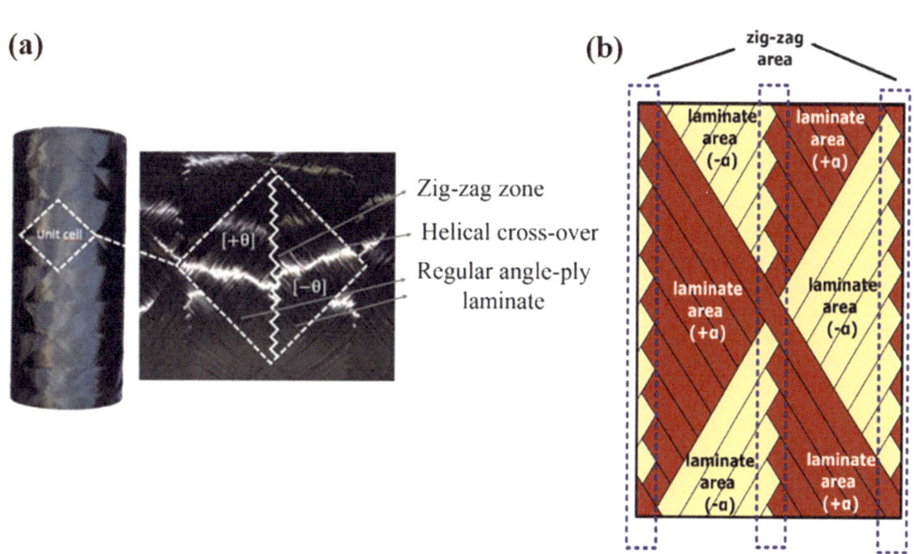

Figure 9.
(a) Winding pattern architecture on composite tubes. (b) Laminate structure – pattern 2/1 with 18 bands (adapted from Ref. [24]).

modifiable resin formulations to meet specific requirements. Recent enhancements in 700 bar type 4 containers have shown promising results, including increased cycling resistance, burst pressure, hydrogen tightness, and improved storage capabilities (**Figure 9**) [28].

5.2 Mozaic pattern

The last decades have seen a more sophisticated process of deposition of the filament settled. Thanks to robotizing, filament winding combines accurate fiber lay-up with a high degree of automation. When filament-wound parts are manufactured, obtaining a pattern geometry is unavoidable [27]. The cyclic positioning of the fiber band on the rotating mandrel creates the so-called mosaic pattern. An integer number defines this pattern, indicating how many diamonds are on the circumference of the part. The diamond regions can then be divided into different areas for analysis. First, two triangular laminate areas with laminate sequences $-\alpha$ / $+\alpha$ and $+\alpha$ / $-\alpha$ can be observed 9. A concentration of interweaving can be seen in this area. It is then concluded that the pattern number directly influences the number of interweaving and undulation areas [29]. A schematic presentation is provided to better understand the creation of interlaces. Finally, the winding pattern architecture on composite tubes is presented. Literature [29–31] suggests that filament winding parameters have been studied in various configurations. The mechanical response of the composite of a ring shape was evaluated to assess the effects of winding angle, diameter-to-thickness ratio, and stacking sequence. A novel hoop ring test facility was used to find fiber and burst properties [12]. Another study investigated impact damage development in composite pipes stacked at different angles. It was found [32] that internal pressure was effective in damaged pipes, and pressure increase reduced impact damage. Additionally, the influence of winding angle on fatigue damage development was studied in glass-fiber-reinforced polymer (GFRP) pipes. The most common failure mode was delamination with small off-axis cracks and fiber/matrix debonding [31]. In this research [32], a genetic algorithm was used to find the best stacking sequence for internally pressurized filament-wound tubes. A novel damage model was developed to predict the response of filament-wound pipes under radial compression and external pressure [33].

An ongoing researcher is still interested in the impact of mosaic patterns on progressive failure. Studies examined pressure tests on glass/epoxy pipe specimens with a ± 55° and ± 75° winding angle [34, 35]. It was concluded that higher pattern numbers, indicating higher degrees of interweaving, influence damage growth. However, this only impacts closed-ended internal pressure loading and weeping tests, with no significant variations found in tensile or pure internal pressure tests. Subsequent significant work was done on pattern issues. Experiments were accompanied by numerical evaluations [36, 37]. Finally [38], an attempt was made to incorporate the mosaic pattern into the cylindrical part and dome area of the numerical investigation.

5.3 Fiber under tension

The preload from pre-tensioning is crucial for enhancing pressure capacity and reducing tank weight and volume [39]. Maintaining this preload at a specific level is important due to varying friction forces between fibers and the mandrel. Higher fiber tension improves rigidity and resilience, while lower tension allows for greater flexibility. Filament-wound composite pressure vessels are designed to stack fibers

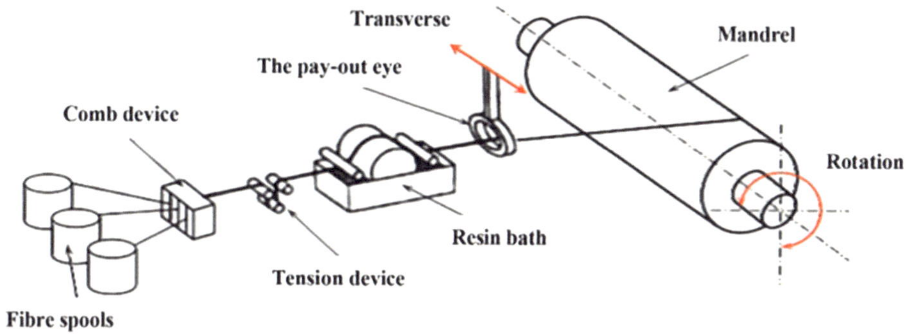

Figure 10.
Schematic diagram of filament winding technique (adapted from Ref. [41]).

with high tension for high-performance applications, typically reaching 60–70% of the fiber's ultimate strength. However, excessive fiber tension can lead to significant elongations of the composite and severe resin matrix splitting between fibers. Resin crazing, occurring between 10% and 40% of final fiber strength, is critical to composite stress, yet it remains lower than operating stress in high-performance CSV. Experimental studies [40] have shown that tubular part strength depends on fiber stress levels, with higher winding tension providing better resistance against failure under fiber-dominated loading conditions. Conversely, reduced fiber tension delays failure under matrix-dominated loading. Additionally, the filament winding process is highly customizable and ideal for automation, particularly in controlling fiber tension and stress.

5.4 Fiber pre-impregnation

In recent years, the use of pre-impregnated fibers (thermoset or thermoplastic) has been introduced to enhance quality control and increase adhesion between laminated ply. Before wrapping around the mandrel, the fibers are soaked in resin (see Illustration 10) and then solidify together with the fiber. Once the fiber wrapping process is complete, the entire assembly, consisting of the mandrel and the layers of composite overwrap, is placed in an oven and heated to the necessary temperatures for curing (**Figure 10**).

Filament-wound composite cylinders exhibit notable advantages with wet winding compared to dry winding. These advantages include reduced material costs, shortened winding duration, and a resin formulation adaptable to specific requirements [42]. Additionally, wet winding offers superior fiber volume control.

6. End-of-life: recycling

The recycling of composite hydrogen storage vessels is crucial because they could offer a more eco-friendly option compared to traditional fuel storage methods. Yet, recycling these vessels can be tricky due to their complex composition and the presence of hydrogen. Recent research [43] has explored various recycling techniques, for instance [44, 45], a mechanical method like grinding and shredding to break down the composite material into smaller pieces, which can then be processed to recover the fibers and resin. Other studies have focused on thermal treatments, such as pyrolysis, to break down the composite material into its basic components. Besides physical recycling methods, there have been investigations into reusing composite hydrogen storage vessels by repurposing them for different applications or refurbishing and refilling them with hydrogen. However, ensuring the integrity and safety of these vessels remains a key consideration for these methods.

In addition to traditional fibers, sustainable and natural fibers have been explored for composite pressure vessels. Bouvier et al. [46] investigated alternative fiber choices for type IV CSVs, considering factors such as mechanical performance, cost, and recyclability. They found that hybrid combinations like E-glass/T700S carbon offer cost advantages, while basalt/recycled T700S and flax/recycled T700S hybrids show potential for reducing greenhouse gas emissions in different pressure vessel applications. However, vessels entirely made of T700S carbon fiber remain the preferred option for maximum mechanical performance.

7. Conclusion

In conclusion, the development and implementation of composite materials, particularly composite fiber-reinforced polymers (CFRP), have revolutionized the field of hydrogen storage vessel manufacturing. These materials, often incorporating high-performance fibers such as carbon fiber, offer a lightweight, high-strength, and corrosion-resistant alternative to conventional storage methods. This review has outlined the critical features of high-performance hydrogen storage vessels, with a focus on the type 4–5 vessels commonly used in various applications.

The primary advantages of composite storage vessels, including their lightweight nature and high strength-to-weight ratio, have been elucidated. These vessels, constructed with a composite shell, inner liner, boss fittings, and carbon filaments, exhibit superior performance in terms of both mechanical strength and gas containment capabilities. Understanding the physical properties and performance metrics of these vessels, such as pressure and volume specifications, is crucial for their successful implementation in hydrogen storage applications.

Furthermore, the choice of epoxy resin as a matrix material for these vessels has been discussed, emphasizing its importance in ensuring the structural integrity and durability of the composite structure. The synthesis and application of carbon fibers in filament winding processes have also been explored, highlighting their role in enhancing mechanical properties and structural performance.

Lastly, the review has addressed the critical issue of end-of-life management for composite storage vessels. Recycling techniques, including mechanical grinding, shredding, and thermal treatments, have been investigated to recover valuable materials such as fibers and resin. Additionally, alternative fiber choices, including sustainable and natural fibers, have been explored for their potential environmental benefits.

Composite hydrogen storage vessels represent a promising solution for the storage and transportation of hydrogen fuel, offering significant advantages in terms of performance, durability, and environmental impact. Continued research and development efforts in this field are essential to further optimize the design, manufacturing, and recycling processes of these vessels, ensuring their continued contribution to the advancement of hydrogen energy technologies.

Author details

Kheireddin Kadri*, Abir Ben Abdallah and Sébastien Ballut
R&D Department, Aptiskills, Levallois-Perret, France

*Address all correspondence to: kheireddin.kadri@aptiskills.fr

References

[1] Pellow MA, Emmott CJ, Barnhart CJ, Benson SM. Hydrogen or batteries for grid storage? A net energy analysis. Energy and Environmental Science. 2015;**8**:1938-1952. DOI: 10.1039/c4ee04041d

[2] Villalonga S, Mathis H, Magnier C. Avancées technologiques pour les réservoirs de véhicules électriques à hydrogène, matériaux et procédés: Un savoirfaire spécifique. In: Chocs Revue Scientifique Et Technique De La Direction Des Applications Militaires. Vol. 51. Villeneuve-Saint-George; 2021. pp. 37-43. Available from: https://www-physique-chimie.cea.fr/physico-chimie-des-materiaux/docs/4/2021/art06_Av_2021.pdf

[3] Makridis SS. Hydrogen storage and compression. In: Carriveau R, Ting DS-K, editors. Methane and Hydrogen for Energy Storage. 1st ed. IET Digital Library; 2016. pp. 1-28. DOI: 10.1049/PBPO101E_ch1

[4] Uetsuji Y, Yasuda S, Teramoto Y. Effect of fibre aspect ratio and aggregation on nonlinear material property of random cellulose reinforced composites: A multiscale computational study. Composite Structures. 2022;**301**:116201. DOI: 10.1016/j.compstruct.2022.116201

[5] Cesano F, Uddin MJ, Lozano K, Zanetti M, Scarano D. All-carbon conductors for electronic and electrical wiring applications. Frontiers in Materials. 2020;**7**:219. DOI: 10.3389/fmats.2020.00219

[6] Humpënoder J. Gas permeation of fibre reinforced plastics. Cryogenics. 1998;**38**:143-147. DOI: 10.1016/S0011-2275(97)00125-2

[7] Flaconnèche B, Martin J, Klopffer MH. Permeability, diffusion and solubility of gases in polyethylene, polyamide 11 and poly(vinylidene fluoride) list of the used notations polymer characteristics. Oil Gas Science and Technology-Rev. IFP. 2001;**56**:261-278

[8] Fujiwara H, Ono H, Ohyama K, Kasai M, Kaneko F, Nishimura S. Hydrogen permeation under high pressure conditions and the destruction of exposed polyethylene-property of polymeric materials for high-pressure hydrogen devices. International Journal of Hydrogen Energy. 2021;**46**:11832-11848. DOI: 10.1016/j.ijhydene.2020.12.223

[9] Condé-Wolter J, Ruf MG, Liebsch A, Lebelt T, Koch I, Drechsler K, et al. Hydrogen permeability of thermoplastic composites and liner systems for future mobility applications. Composites Part A: Applied Science and Manufacturing. 2023;**167**:107446. DOI: 10.1016/j.compositesa.2023.107446

[10] Ebermann M, Bogenfeld R, Kreikemeier J, Glüge R. Analytical and numerical approach to determine effective diffusion coefficients for composite pressure vessels. Composite Structures. 2022;**291**:115616. DOI: 10.1016/j.compstruct.2022.115616

[11] Air A, Shamsuddoha M, Gangadhara Prusty B. A review of type V composite pressure vessels and automated fibre placement based manufacturing. Composites Part B: Engineering. 2023;**253**:110573. DOI: 10.1016/j.compositesb.2023.110573

[12] Kumazawa H, Susuki I, Aoki T. Gas leakage evaluation of cfrp

cross-ply laminates under biaxial loadings. Journal of Composite Materials. 2006;**40**:853-871. DOI: 10.1177/0021998305056379

[13] Tomohiro Yokozeki TA, Ishikawa T. The effect of matrix cracks on gas permeability through cfrp laminates. Advanced Composite Materials. 2004;**13**:227-236. DOI: 10.1163/1568551042580136

[14] Schäkel M, Janssen H, Brecher C. Process analysis of manufacturing thermoplastic type-iv composite pressure vessels with helical winding pattern. In: Society for the Advancement of Material and Process Engineering (SAMPE International Conference) 2021. 2021. pp. 955-967. DOI: 10.24406/publica-fhg-417015. Available from: https://publica.fraunhofer.de/handle/publica/417015

[15] Zhang J, Yan Y, Zhang C, Xu Z, Li X, Zhao G, et al. Properties improvement of composite layer of cryo-compressed hydrogen storage vessel by polyethylene glycol modified epoxy resin. International Journal of Hydrogen Energy. 2022;**48**:5576-5594. DOI: 10.1016/j.ijhydene.2022.11.138

[16] Hwang T-K, Hong C-S, Kim C-G. Size effect on the fiber strength of composite pressure vessels. Composite Structures. 2003;**59**:489-498

[17] Cohen D, Mantell SC, Zhao L. The effect of fiber volume fraction on filament wound composite pressure vessel strength. Composites: Part B. 2001;**32**:413-429

[18] Newcomb BA. Processing, structure, and properties of carbon fibers. Composites Part A Applied Science and Manufacturing. 2016;**91**:262-282. DOI: 10.1016/j.compositesa.2016.10.018

[19] Zhao C, Li X, Wang L, Zhang Y. Research on mechanical properties of t700 carbon fiber under high temperature. IOP Conference Series: Materials Science and Engineering. 2017;**234**:052021. DOI: 10.1088/1757-899X/234/5/052021

[20] Li M, Lian Z, Lai H, Hu G. Processing and tensile properties of continuous carbon fiber m40 composite. Key Engineering Materials. 2016;**703**:123-128. DOI: 10.4028/www.scientific.net/KEM.703.123

[21] Mao Y, Chen D, Thomason J, Brown A. Mechanical properties of im7/977-2 composite at elevated temperature and in a marine environment. Polymer Composites. 2016;**37**:587-596. DOI: 10.1002/pc.23341

[22] Cao Y, Guo J, Hou J, Huang Y. Effect of thermal treatment on mechanical properties of t800 carbon fiber. Materials Research Express. 2019;**6**:126527. DOI: 10.1088/2053-1591/ab5dc4

[23] Bai Y, Li B, Hu C, Wu Y, Geng J. Study on interfacial properties and mechanical performance of torayca® t800s carbon fiber reinforced epoxy composites. Key Engineering Materials. 2017;**727**:171-176. DOI: 10.4028/www.scientific.net/KEM.727.171

[24] Azeem M, Ya HH, Kumar M, Stabla P, Smolnicki M, Gemi L, et al. Application of filament winding technology in composite pressure vessels and challenges: A review. Journal of Energy Storage. 2022;**49**:103468. DOI: 10.1016/j.est.2021.103468

[25] Zhou W, Wang J, Pan Z-b, Liu J, Ma L-h, Zhou J-y, et al. Review on optimization design, failure analysis and non-destructive testing of composite hydrogen storage vessel. International Journal of Hydrogen Energy.

2022;**47**:38862-38883. DOI: 10.1016/j.ijhydene.2022.09.028

[26] Sulaiman S, Borazjani S, Tang SH. Finite element analysis of filament-wound composite pressure vessel under internal pressure. IOP Conference Series: Materials Science and Engineering. 2013;**50**:012061. DOI: 10.1088/1757-899X/50/1/012061

[27] Dalibor IH, Lisboa TV, Marczak RJ, Amico SC. A geometric approach for filament winding pattern generation and study of the influence of the slippage coefficient. Journal of the Brazilian Society of Mechanical Sciences and Engineering. 2019;**41**:1-16. DOI: 10.1007/s40430-019-2083-2

[28] Berro Ramirez JP, Halm D, Grandidier JC, Villalonga S, Nony F. 700 bar type IV high pressure hydrogen storage vessel burst - simulation and experimental validation. International Journal of Hydrogen Energy. 2015;**40**:13183-13192. DOI: 10.1016/j.ijhydene.2015.05.126

[29] Eggers F, Humberto J, Jr SA, Azevedo CB, Amico SC. Mechanical response of filament wound composite rings under tension and compression. Polymer Testing. 2019;**78**:105951. DOI: 10.1016/j.polymertesting.2019.105951

[30] Hwang T-K, Park J-B, Kim H-G. Evaluation of fiber material properties in filament-wound composite pressure vessels. Composites: Part A. 2012;**43**:1467-1475. DOI: 10.1016/j.compositesa.2012.04.005

[31] Yazman S. The effects of back-up on drilling machinability of filament wound GFRP composite pipes: Mechanical characterization and drilling tests. Journal of Manufacturing Processes. 2021;**68**:1535-1552. DOI: 10.1016/j.jmapro.2021.06.054

[32] Humberto J, Jr SA, Ribeiro ML, Tita V, Amico SC. Stacking sequence optimization in composite tubes under internal pressure based on genetic algorithm accounting for progressive damage. Composite Structures. 2017;**178**:20-26. DOI: 10.1016/j.compstruct.2017.07.054

[33] Xia M, Takayanagi H, Kemmochi K. Analysis of multi-layered filament-wound composite pipes under internal pressure. Composite Structures. 2001;**53**:483-491

[34] Gemi L, Yapici A. Science and fatigue. Composites Science and Technology. 2005;**65**:703-708. DOI: 10.1016/j.compscitech.2004.10.002

[35] Akdemir A, Şahin ÖS. Progressive fatigue failure behavior of glass/epoxy (± 75) 2 filament-wound pipes under pure internal pressure. Materials & Design. 2009;**30**:4293-4298. DOI: 10.1016/j.matdes.2009.04.025

[36] Douchin B. Influence of winding pattern on the mechanical behavior of filament wound composite cylinders under external pressure. Composites Science and Technology. 2008;**68**:1015-1024. DOI: 10.1016/j.compscitech.2007.07.020

[37] Shen C, Han X. Damage and failure analysis of filament wound composite structure considering fibre crossover and undulation. Advanced Composites Letters. 2018;**27**(2). DOI: 10.1177/096369351802700202

[38] Stabla P, Smolnicki M, Blazejewski W. The numerical approach to mosaic patterns in filament - wound composite pipes. Applied Composite Materials. 2021;**28**:181-199. DOI: 10.1007/s10443-020-09861-z

[39] Drozdov AD. Optimization of winding process for composite pressure

vessels. International Journal of Pressure Vessels and Piping. 1995;**62**:69-81

[40] Mertiny P, Ellyin F. Influence of the filament winding tension on physical and mechanical properties of reinforced composites. Composites: Part A. 2002;**33**:1615-1622

[41] Manojprabhakar M, Rajini N, Mayandi K. Wireless technology applied in 3-axis filament winding machine control system using MIT app inventor wireless technology applied in 3-axis filament winding machine control system using MIT app inventor. IOP Conference Series: Materials Science and Engineering. 2019;**469**:012030. DOI: 10.1088/1757-899X/469/1/012030

[42] Wang Q, Zhang Q, Kang Y, Wang Y, Liu J. An investigation of preparation of continuous carbon fiber reinforced PLA prepreg filament. Composites Communications. 2023;**39**:101530. DOI: 10.1016/j.coco.2023.101530

[43] Bernatas R, Dagreou S, Despax-ferreres A, Barasinski A. Recycling of fiber reinforced composites with a focus on thermoplastic composites. Cleaner Engineering and Technology. 2021;**5**:100272. DOI: 10.1016/j.clet.2021.100272

[44] Nachtane M, Tarfaoui M, Abichou MA, Vetcher A, Rouway M, Aâmir A, et al. An overview of the recent advances in composite materials and artificial intelligence for hydrogen storage vessels design. Journal of Composites Science. 2023;7(3):119

[45] Zhang J, Chevali VS, Wang H, Wang CH. Current status of carbon fibre and carbon fibre composites recycling, Composites Part B Engineering. 2020;**193**:108053. DOI: 10.1016/j.compositesb.2020.108053

[46] Bouvier M, Guiheneuf V, Jean-marie A. Modeling and simulation of a composite high-pressure vessel made of sustainable and renewable alternative fibers. International Journal of Hydrogen Energy. 2019;**44**:11970-11978. DOI: 10.1016/j.ijhydene.2019.03.088

Chapter 4

Hydrogen Storage Employing Select, Main-Group-Based Inorganic Materials

Steven Snow, Trisha Hoover and Malcolm Penman

Abstract

The use of hydrogen as a fuel is considered a serious option to reduce the long-term environmental impact of global warming. A key challenge of using hydrogen as a fuel is that of employing safe and effective methods by which to store it. One general approach to addressing this challenge is to utilize chemical compounds that release hydrogen gas under highly specified and controlled chemical conditions. This review will discuss said compounds which contain selected main-group inorganic elements, including certain (1) Alkaline-based metals (Li, Na, K, Mg, Ca), (2) Boron and Aluminum, and (3) Silicon. The majority of these compounds release hydrogen gas under mild conditions, typically by hydrolysis. The performance criteria of these compounds will be compared along with commentary on the topics of (1) Synthesis of these materials, (2) Energy requirements, (3) Hydrogen release chemistry, (4) Handling safety, and (5) The challenges of recycling/ reloading these materials.

Keywords: storage, chemical release, main group elements, hydride, protonic, hydrolysis, terrestrial, handling safety

1. Introduction

The concept of global warming due to the atmospheric build-up of carbon dioxide via the combustion of fossil fuels is firmly established in the scientific community and other spheres that strongly influence energy policy-making [1]. One of the most intensely investigated alternatives to fossil fuels is that of molecular hydrogen, H_2 [1–6].

Concerned stakeholders generally agree that a major drawback in the use of hydrogen as an energy source is the lack of methods available to effectively store it [7–15]. Regarding *storage*, although hydrogen gas has a favorably high weight energy density (120 MJ/kg at 25°C and atmospheric pressure) [16], that advantage is compromised by its unfavorably low volumetric energy density (10.05 kJ/L at 25°C and atmospheric pressure) [16]. Therefore, the current mature storage technology is H_2 gas compression at 70 MPa. (resulting in a volumetric energy density of approximately 3000 kJ/L) or else cryo-compression of the gas to a liquid, at the exceptionally low temperature of 20 deg. K. Unfortunately, these storage methods are hampered by the considerations that, under these conditions; (1) Many metals used in storage tanks become embrittled, (2) Handling either a high-pressure gas or else an exceptionally cold, cryogenic liquid[1] may not compatible with mass-market applications where untrained individuals are involved (such as hydrogen fuel cell-powered vehicles), and (3) Hydrogen is explosively flammable in air under relatively easy-to-achieve conditions in a closed environment (see Footnote #1). These afore-mentioned challenges spurred an intense effort to develop alternative methods of storing H_2.

Generally, when one considers "alternative" storage systems for H_2, the working concept is that of substituting "pure" H_2 within the storage container with a material that releases H_2 at a controlled rate under specific operational conditions. For example, one can consider sodium metal to be a material that fits these criteria. Under typical storage conditions (near room temperature and atmospheric pressure), when protected from moisture, sodium metal is sufficiently stable to be utilized in storage applications. Furthermore, when one seeks to "release" H_2 from sodium, this can be accomplished by exposing the metal to a chemical that contains what is commonly referred to as an "active" (acidic) proton. Common chemicals that contain "active" protons include water, alcohols, silanols, carboxylic acids, ammonia and certain amines, etc. The chemical reaction between sodium and, for example, water, is relatively rapid, highly exothermic and yields hydrogen gas and sodium hydroxide as products.

$$Na\ (s) + H_2O \rightarrow NaOH\ (s) + \tfrac{1}{2}\, H_2\ (g) \qquad (1)$$

The formation of a by-product, NaOH, hints at another common feature of many (but not all) of these types of H_2 storage systems, that being the need to remove a chemical by-product from the storage system and replace it with more storage material, in this case, sodium metal.[2]

Now that we have established that "alternative" H_2 storage methodologies involve the "release" or "generation" of H_2 from a defined, stored material, it is convenient to categorize the commonly referred to types of storage *processes*:

1 When hydrogen is stored as liquid at 1 atm, it must be maintained below its very low boiling point. Despite displaying an improved volumetric energy density versus compressed gas storage, liquid hydrogen (LH_2) storage is not frequently used for several reasons. One reason is that, with its exceptionally low boiling point, even stored within a highly insulated tank, the LH_2 can evaporate, which causes pressure build-up leading to the necessity, to prevent rupture, to have tank venting to the atmosphere. This results in hydrogen loss. Apart from the cost and energy efficiency penalties suffered, the "boil-off" presents safety concerns, particularly when the gas release occurs in confined spaces such as in parking garages (for H_2 gas, at room temperature and atmospheric pressure, while exposed to an appropriately strong electric spark, the flammability range is 4–75% and the detonation range is 18.3–59%) [17]. Secondly, 30–35% of energy value of hydrogen is required to liquefy it, which is three times larger than the energy required to compress gaseous hydrogen for storage. For these reasons, even though there is room for further improvements, it is unlikely for cryogenic storage to meet US Department of Energy (DOE) requirements. For an overview of safety concerns handling liquid hydrogen see [18, 19].

2 The presence of by-products is a defining feature of chemical forms of energy. Even fossil fuels have by-products of chemical combustion, specifically water vapor and carbon dioxide, which, unlike the example of sodium and water yielding NaOH as a by-product, conveniently remove themselves as gaseous exhausts from the engine.

1. *Physisorption, Storage, and Desorption ("Release") of H_2-* Reversible adsorption of H_2 on a solid or liquid where the H-H bond in H_2 is preserved throughout the entire process. The "adsorbing" material is usually not altered due to the process and can typically be reused.

2. *Chemisorption, Storage, and Desorption of H_2-* Reversible adsorption of H_2 on a solid or liquid where the H-H bond in H_2 is broken during the adsorption process, the "breakage" is maintained through the "storage" process, and then is reformed during the desorption ("release") process. The "adsorbing" material may or may not altered due to the process and can, in favorable cases, be reused.

3. *The storage of a material which may, or may not, contain hydrogen atoms, followed by a specifically-defined, chemical process which results in the controlled formation of H_2 gas. In the case where the stored material does not contain hydrogen atoms, then the stored material is exposed to a 2nd material which does contain the hydrogen atoms which ultimately form H_2-* During this process there is typically a substantial degree of breakage and reformation of chemical bonds, highlighted by the breakage of X-H bonds (where X can be virtually any element in the periodic table) and the subsequent formation of H-H bonds. Along with the generation of H_2 gas, chemical by-products are formed that need to be ultimately removed from the storage container. An example of this process is shown in Eq. (1).

This chapter will focus on examples of Process #3 listed above. The primary reason for this focus is that materials undergoing Process # 3 potentially produce more H_2 per storage/release cycle than do the materials that undergo Processes #1 and #2 [13, 20]. Within the broad range of materials and processes that meet the definition of Process #3, this chapter will focus on storage materials where the central element X of the material is referred to as a "main group" element of the Periodic Table of the Elements (Groups # 1, 2, 13–18)[3] [21]. Furthermore, within the "main group" elements, we will narrow our scope of review to those groups where, for the element X, the X-H bond is *hydridic* in nature. A "hydridic" X-H bond is one where the electron distribution in the bond is unbalanced with the majority of it located near the hydrogen atom. Main-group compounds with hydridic character constitute the majority of examples associated with the H_2 storage application.[4] Therefore, this chapter will cover materials which contain chemical elements from Groups # 1, 2, 13 & 14 of the Periodic Table which also meet certain criteria.[5] The result of subjecting the Periodic Table to

---

[3] A substantial literature exists regarding the utilization of compounds in the H_2 storage application containing X-H bonds where X is either a transition metal or else a rare-earth element. However, there appears to be much more literature covering the main group elements; therefore, the authors have chosen to focus on them.

[4] A key exception to this generalization is the case of ammonia, NH_3, which has a protonic N-H bond, where the electron density is localized at the nitrogen atom. A significant literature exists discussing the utility of ammonia in the H_2 storage application, for example [22].

[5] For two reasons, certain elements within these groups will not be covered in this chapter. These reasons are (1) The element must have a sufficiently high terrestrial concentration where it can be economically harvested and refined, and (2) It cannot be either radioactive or exceptionally toxic. These criteria eliminate the following elements from further consideration [Group 1- Rb, Cs, Fr; Group 2-Be, Sr., Ba, Ra; Group 13-Ga, In, Tl; Group 14-Ge, Pb.

these limitations leaves us with seven elements: Lithium (Li), Sodium (Na), Potassium (K), Magnesium (Mg), Boron (B), Aluminum (Al), and Silicon (Si).

2. Main group element-based hydrogen storage materials (MGE-HSMs)

2.1 General trends relevant to the H_2 storage application

2.1.1 Hydridic X-H bonds react with protonic X-H bonds to form hydrogen

Regarding the seven elements selected to be further reviewed, for the great majority of them, the materials which find applicability for H_2 storage contain one or more hydridic X-H bonds; in fact, the hydridic bond is usually the source of generation of H_2 gas. The polarity of X-H bonds within the materials covered in this chapter range from X^+H^- (where the electron charge distribution is so localized at the H atom that the bond is "ionic"), through a range of X-H bond polarities where the bond has substantial covalent character.

Regarding the H_2 storage application, the key working chemistry principle is that: "Hydridic X-H bonds react with Protonic X-H bonds to form H_2". A "protonic" X-H bond is the case where the hydrogen atom in the bond has a degree of positive charge. A relevant and highly common example is given in Eq. (2). In this case, the hydridic X-H bond exists in NaH[6] and the protonic X-H bond exists in H_2O.

$$NaH\ (s) + H_2O \rightarrow NaOH\ (s) + H_2\ (g) \qquad (2)$$

To extend this unifying concept even further, the polarity of X-H bonds, all else being equal, correlates to the position of an element within the Periodic Table of the Elements [23]. Regarding this correlation, one can apply the following "rules of thumb": (1) As one descends down a Group (or alternatively a "column") in the Table the X-H bond becomes more hydridic (therefore, the K-H bond in potassium hydride is more hydridic than the Na-H bond in sodium hydride), and (2) As one moves from right-to-left along a row in the Table, the X-H bond becomes more hydridic (therefore, the Mg-H bond in magnesium hydride is more hydridic than the B-H bond in boron hydrides).

Regarding the H_2 storage application, one can apply the "rules" above in order to make qualitative predictions about the chemistries required to yield H_2 gas under various conditions. To wit, regarding the uppermost row in The Periodic Table (Na, Mg, B, C, N, O, F), one might expect the following outcomes: (1) In order to harvest H_2 from either NaH or MgH_2, both highly hydridic, one would use a reagent with a protonic X-H bond, and a weakly protonic agent such as ammonia, NH_3, (2) The boron hydrides are less hydridic than either NaH or MgH_2; therefore, a reagent that is moderately protonic (such as water or, if necessary, acetic acid, CH_3COOH)[7] might be needed to efficiently release H_2, (3) The C-H bond, particularly in molecules not containing many electronegative atoms (typically nitrogen or oxygen), is close to being non-polar and, therefore, is quite resistant to the reaction with either hydridic

---

[6] The hydridic nature of the Na-H bond is the reason that the compound is called "sodium hydride".

[7] As might be expected, there is a significant correlation between the rate of H_2 generation from hydridic bonds and the strength of the acidity (pKa) of the protonic reactant.

or protonic X-H bonds to form H_2,[8] and (4) Finally, the N-H, O-H and F-H bonds are progressively more protonic, and therefore require hydridic X-H reagents, such as NaH, to release H_2 gas.

The concepts developed above can be generalized to expand the "chemistry set" available to produce H_2 gas. For example, returning to Eq. (2) above, one can also categorize this reaction as a reduction/oxidation ("redox") process. Specifically, water can be the "oxidizing" agent, formally removing an electron from the hydride (H^-) ion, yielding the hydrogen atom, which rapidly dimerizes to form H_2. Alternatively, one can describe this process as the reduction of water (by the hydride "reducing agent"), yielding the hydroxide (OH^-) ion (as NaOH). The utility of carrying out this categorization is that one can conceive of a much wider range of redox chemistry which can potentially yield H_2 gas "on demand". For example, in Eq. (3), it is problematic to consider the chemical reaction as one between a "hydridic" compound and a "protonic" compound to yield H_2; specifically there is no "hydridic" compound present as a reagent.

$$Na\ (s) + H_2O \rightarrow NaOH\ (s) + \tfrac{1}{2}\, H_2\ (g) \tag{3}$$

Equation 3 is more accurately described as a "redox" reaction. Sodium metal has an extensive chemical portfolio as a strong reducing agent [21]. Within Eq. (3), sodium reduces water,[9] yielding the reaction products including H_2 gas.

2.1.2 The terrestrial forms of MGE-HSMs

It is well-known that carbon-centered "fossil fuels" were created geochemically, over the millennia, due to the decomposition of animal and plant residues in anaerobic conditions [25]. The anaerobic conditions were necessary for the formation of their signature C-C and C-H bonds. Under aerobic conditions, in the presence of an ignition source, these bonds will rapidly react with oxygen, forming C-O and O-H bonds. These are the essential steps in the conventional energy-creating, fossil fuel-based, combustion process. In the absence of an ignition source, under conditions commonly found on the surface of the earth, fossil fuels have significant chemical stability.[10] Furthermore, physical forms of the element carbon existing under terrestrial conditions,[11] such as graphite, are quite stable to chemical change.

---

[8] The major ingredients of fossil fuels are the straight chain and branched chain hydrocarbons such as octane, which contain only non-polar C-C bonds and minimally polar C-H bonds. Therefore, one would hypothesize that these compounds would be highly resistant to reacting with either hydridic, or else protonic, compounds to yield H_2 gas. This turns out to be the case. In one reported example, in order to harvest H_2 from hydrocarbons, one must use extremely strong acids ("superacids") under fairly aggressive conditions (140 deg. C) [24].

[9] An alternative, and possibly more descriptive interpretation of this chemistry is that "Sodium metal reduces (donates an electron to) the protonic hydrogen on water, yielding the hydrogen atom, which rapidly dimerizes to form H_2".

[10] One of the primary reasons that they are universally utilized as an energy source.

[11] Strictly for the purpose of this article, the term "*terrestrial conditions*" will refer to the conditions (often within the context of "chemistry") present in the atmosphere, surface and subterranean aspect (where mining operations can be carried out) of the Earth. Therefore, the *terrestrial "state" or "form" of an element* is its material composition and structure under terrestrial conditions. Separately, the term *"native state" of the element* is used in this document. This term specifically refers to the molecular structure of the element in the absence of any other elements; for example, "pure" silicon metal. With a number of significant exceptions, including graphitic carbon, typically, the "terrestrial" and "native" states of an element are not the same.

Unlike the example of carbon discussed above, many of the main group elements[12] considered in this chapter, in forms relevant to the H_2 storage application (MGE-HSM), whether in compounds containing X-H bonds or else the element in its native state,[13] are *not* chemically stable under terrestrial conditions. These elements are usually harvested as highly-oxygenated compounds such as the oxides, hydroxides and silicates. This is consistent with an earth atmosphere whose most reactive gases are oxygen and water. The chloride ion is also quite abundant in the earth's crust, resulting in the presence of the chloride forms of many of these elements.

2.1.3 Storage and reaction

When one considers the MGE-HSMs within this chapter, in contrast to the case of fossil fuels where the stored fuel (for example, that in one's gas tank) is directly combusted to yield the energy content, it is necessary to carry out a chemical process on the stored material in order to yield the combustible H_2. This is the necessary "trade-off" for utilizing a storage material that is, by choice, easier to handle than either gaseous or liquid H_2.

Therefore, when one considers utilizing these MGE-HSMs, they have two "high-level" choices: (1) Transport the MGE-HSM from the storage container to a separate reactor compartment to carry out the H_2 release chemistry, or (2) Carry out the said chemical process within the "storage" container.

2.2 Chemical reactivity patterns

2.2.1 Homogeneous (single phase) reactions

Assuming that (1) The H_2 generation process occurs within a single phase, and (2) The rate of the reaction is not reagent-diffusion-rate-controlled; then the rate of reaction between an MGE-HSM and its "activator" (reagent used to generate H_2 gas) will often be a second-order rate process with rate dependencies on the concentration of both the MGE-HSM and its activator. One result of this set of initial conditions is, if the MGE-HSM and its activator are mixed together with a mutual solvent, and none of either reagent is added to the mixture afterwards, then the rate of evolution of H_2 gas will exponentially decrease with time. Assuming that under operational conditions (such as supplying H_2 gas to a working fuel cell) one desires *a constant rate of evolution* of H_2 gas, one would need to implement either mechanical and/or chemical modifications to the storage/reactor/gas transfer assembly. For example, in terms of a chemical modification, one could potentially control the rate of introduction of activator into the storage/reactor in such a method that the *product* of the two reagent concentrations remain constant throughout the course of the reaction.

---

[12] Regarding the seven elements that we have chosen to review, the (partial) exceptions to this generalization are boron (B) and silicon (Si). Certain compounds containing B-H and Si-H bonds are quite stable under terrestrial conditions as are native silicon and boron in the cast form (for example, silicon "metal"). Furthermore, casted silicon has many large-scale "terrestrial" applications such as photovoltaic solar cells [26–28]. On the other hand silane (SiH_4), the structural analog to carbon-based methane, and diborane (B_2H_6), are both spontaneously inflammable in air.

[13] See Footnote # 11.

2.2.2 Heterogeneous (multiple phase) reactions

Assuming that (1) The H_2 generation process occurs in a multiple phase system, and (2) The rate of the reaction is not reagent-diffusion-rate-controlled, the rate of reaction between an MSG-HMS and it's "activator" will, along with potential rate dependencies on the concentration of both reagents, will potentially have rate dependencies on the *total interfacial area* where reaction occurs. In the context of MSG-HMS chemistry, a common example of this situation is one where the MSG-HMS is being stored as a solid and the activator is introduced into the storage compartment as a liquid or a vapor. The H_2-generating reaction occurs at the surface of the solid MSG-HMS. While the rate of the reaction may depend upon the amounts of MGE-HMS and activator, it will also depend on the total surface area of the solid MGE-HMS. Furthermore, for a given amount of solid, the total surface area will be inversely proportional to the particle size (overall, smaller particles ➔ larger surface area ➔ faster reaction). Overall, this is a complex kinetic situation, and establishing a constant rate of H_2 evolution would be a non-trivial technical challenge.

2.2.3 Heat release and exchange

Another common challenge in generating H_2 gas from the "activation" of an MGE-HSM is that often the activation reaction is both rapid and highly exothermic. Although the rate of reaction can be controlled by a plethora of technical options, efficient heat removal from the storage/reaction container is necessary (if nothing else, to safely operate the engine consuming the H_2 gas) and practical utilization of the heat generated would be a highly-desired outcome.

2.2.4 Removal of non-H_2, non-volatile, reaction products

Typically, the generation of H_2 from MGE-HSM materials will yield non-volatile reaction products that will need to be removed from the storage/reaction container at the end of a fueling/combustion cycle.

3. The transformation of terrestrial forms of MGE-HSMs to useful hydrogen storage materials and their subsequent performance

A summary of the relevant MGE-HSMs is displayed in **Table 1** including (1) Their terrestrial forms, (2) The chemical transformations necessary to convert them to storable materials, (3) The chemistry necessary to release H_2 gas from the stored material and, (4) Their relative abundances of their central element in the earth's crust [42].

As displayed in **Table 1**, one general result arising from the harvesting of the terrestrial forms of these specific elements is that often a multiple series of chemical transformations is necessary to yield a useful HSM. Furthermore, as a trend, the chemical reactions necessary to yield the HSM tend to be substantially endothermic with high activation energies [21, 23].

Element	A common terrestrial form of the element	Element transformation to H_2 storage material	A common example of H_2 release chemistry	RAEC[a]
Li	$LiAlSi_2O_6$ (Spodumene)	$LiAlSi_2O_6 \rightarrow LiCl \rightarrow Li(s) \rightarrow LiH$	$LiH + H_2O \rightarrow LiOH + H_2$	$7.09(10^{-5})$
Na	NaCl (Rock Salt)	$NaCl \rightarrow Na(s) \rightarrow NaH$	$NaH + H_2O \rightarrow NaOH + H_2$	0.08
K	$KCl/ MgCl_2/ 6\ H_2O$ (Carnalite)	Carnalite $\rightarrow KCl \rightarrow K \rightarrow KH$	$KH + H_2O \rightarrow KOH + H_2$	0.07
Mg, Ca	$CaCO_3/ MgCO_3$ (Dolomite)	Dolomite $\rightarrow MgO/CaO \rightarrow Mg, Ca \rightarrow MgH_2, CaH_2$	$CaH_2 + H_2O \rightarrow Ca(OH)_2 + H_2$	0.08 (Mg) 0.15 (Ca)
B	B_2O_3 (Boric oxide), $B(OH)_3$ (Boric acid) and their various polymeric forms	Boric oxide/ Boric acid $\rightarrow B(OMe)_3 \rightarrow NaBH_4$; Borate/Boric acid $\rightarrow BF_3$; $NaBH_4 + BF_3 \rightarrow B_2H_6$	$NaBH_4 + (2 + x)\ H_2O \rightarrow 4\ H_2 + NaBO_2{\cdot}x\ H_2O$. $B_2H_6 + 3\ H_2O \rightarrow 2\ B(OH)_3 + 3\ H_2$	$3.55(10^{-5})$
Al	$Al(OH)_3$ (Gibbsite- one component of Bauxite)	$Al(OH)_3 \rightarrow Al_2O_3 \rightarrow Al \rightarrow NaAlH_4 \rightarrow LiAlH_4$	$Li[AlH_4] + 4\ H_2O \rightarrow LiOH + Al(OH)_3 + 4\ H_2$	0.29
Si	SiO_2 (Silica)	$SiO_2 \rightarrow Si \rightarrow Si_xH_y$; $SiO_2 \rightarrow SiH_4$	$SiH_4 + O_2 \rightarrow SiO_2 + 2\ H_2$	1

[a]RAEC is an acronym for "Relative Abundance in the Earth's Crust".
Ref. Groups I and II elements—[29, 30].
Ref. Boron—[31–33].
Ref. Aluminum—[34–37].
Ref. Silicon—[26–28, 38–41].

Table 1.
Description of the processes necessary to convert MGE-HSMs into their storage forms and the associated H_2 generation chemical reactions.

3.1 Group I and II elements (Li, Na, K, Mg, Ca)

The MSG-HSMs of Groups I and II of the Periodic Table of the Elements share the characteristic of containing a "hydridic" ionic bond, M^+H^-. Therefore, based on the concepts discussed in Section IIIa, the following "protonic source reactivity" scale can be asserted:

$MgH_2 < CaH_2, LiH < NaH < KH$

For the three Group I elements, the associated hydride is produced from the native state of the element which, itself, is produced from the chloride salt. For Na and K, the chloride salt is the terrestrial form of the element. For Li, the chloride salt is synthesized from the terrestrial mineral Spodumene ($LiAlSi_2O_6$). The hydride salts of all of the Group I elements are crystalline solids at room temperature and are highly reactive, yielding substantial exothermicity, with water. Therefore, with regards to safety, for the H_2 storage application, they need to reside in an H_2O-free, ventable, storage device.

When assessing the effectiveness of an MGE-HSM, it is also useful to consider the terrestrial abundance of that specific element and the on-going efforts being made to harvest it. The relative abundances of these three Group I elements are displayed in **Table 1**. Both Na and K have substantial terrestrial abundances while Li is significantly less. All three of these elements are part of materials that are utilized in large and commercially important applications not directly related to the use of hydrogen as a fuel; therefore, substantial infrastructure exists to harvest these elements in some form. In particular, despite its low natural abundance, during the last few decades there have been expansive worldwide efforts to harvest terrestrial lithium. This effort is being driven by large commercial applications, in particular, the fabrication of lithium-ion batteries.

The situations for magnesium and calcium mirror those in place for the Group I MGE-HSMs.

Overall, the main advantages of utilizing Group I and II MSG-HSMs includes: (1) With the exception of lithium, the terrestrial forms of the elements are in relatively high abundance and, including for lithium, substantial infrastructure is currently in place to harvest them, (2) The hydride compounds of these elements are very reactive to protonic reagents; therefore, many options exist to control the H_2 generation processes, and (3) The by-products of the H_2 generation process, the various hydroxide salts, are all chemicals that have an extensive industrial processing history. The disadvantages of using these MGE-HSMs are (1) Their production requires multi-step, highly endothermic processes regarding substantial energy input and often releasing unwanted CO_2 into the environment, (2) In the presence of water they are unstable, even explosive, which is very problematic, particularly when in the presence of untrained personnel.

3.2 Group XIII and XIV elements (B, Al, Si)

For all three of the above elements, substantial infrastructure exists to harvest their terrestrial compound forms. In common with the Group I and II MGE-HSMs, the M-H bonds in boron (B), aluminum (Al) and silicon (Si) MGE-HSMs are *hydridic*. Therefore, these M-H bonds exhibit substantial reactivity with protonic materials, yielding H_2 gas. However, as expected from an analysis of the Periodic Table, these M-H bonds are not as polar as those of the Groups I/II, and therefore are not as reactive with protonic reagents.

3.2.1 Boron

As displayed in **Table 1**, the terrestrial forms of boron include boric oxide (B_2O_3), boric acid [$B(OH)_3$] and various polymeric forms of these compounds. With regards to the H_2 storage application, it is the hydrides of boron which are the materials of interest. Unlike the Group I and II hydrides, which are essentially encompassed by the ionic hydrides M-H, the boron hydrides (boranes) and their relevant derivatives have a rich structural complexity [43, 44]. For the boranes, electron-deficient bonding leads to exceptional reactivity for the lower boranes (di-, tetra- and pentaborane), including their spontaneous inflammability in air. This reactivity is diminished as either (1) The MW of the borane increases (along with degree of delocalized bonding), or else (2) Base adducts are formed, relieving the degree of electron-deficient bonding (ammonia-boranes, borohydrides, etc.).

Of all the elements discussed in this chapter, MGE-HSMs based on boron have received the most R&D attention. For example, in the 2010–2015-time frame, the US Department of Energy (US-DOE) funded a substantial program to identify the best candidate HSMs, with an emphases on (a) Fuel cell/vehicle transport applications, and (b) The performance metric of direct hydrogenation of the "spent" fuel. The materials screened numbered well into the hundreds [9]. One result of this program was the selection of a boron-containing MGE-HMS, ammonia-borane ($H_3B\text{-}NH_3$) as the top candidate for the H_2 storage application. This selection was the first step in a detailed and comprehensive investigation of this material as an MGE-HSM, with over 300 publications created. Demirci recently published an overview and critique of this large body of work [45].

Ammonia-borane can be classified as a base adduct of "borane" (BH_3), where the "base" is ammonia, NH_3, and the chemical bond between nitrogen and boron is the result of the donation of the lone pair of electrons on the nitrogen atom of ammonia to boron. Base adducts of the boranes are very common and, in the case where the "base" is the hydride ion, H-, the resultant adduct is termed a "borohydride", BH_4^-. Borohydride salts, particularly lithium, sodium and potassium, have been intensively investigated in the H_2 storage application, in part due to, in analogy with the case for $H_3B\text{-}NH_3$, their (1) Relatively high hydrogen densities, and (2) Stability in air [20, 33].

With regard to the potential reversible hydrogenation/dehydrogenation of boron-containing materials, Severa and co-workers demonstrated the direct hydrogenation of magnesium boride (MgB_2), to magnesium borohydride [$Mg(BH_4)_2$], under 950 bar H_2 at 400 deg. C. [46].

To date, to the authors' knowledge, despite all of this R&D work, no boron-centered MSG-HMS have yet been commercialized for the H_2 storage application. The authors conclude that there are two reasons for that: (1) In agreement with Demirci [45], not enough work has been carried out investigating the processes needed to scale-up the most promising materials to the production environment, and (2) The threat of the unintended creation and release of explosive, low molecular weight boranes into the air.

3.2.2 Aluminum

With aluminum (Al) positioned just below boron in the Periodic Table, one would expect, all else equal, the Al-H bond to be more ionic, and therefore, more *hydridic*, than the B-H bond. This should, and does, translate into a higher reactivity with protonic sources, such as water [47].

A number of aluminum hydrides have been investigated as HSMs [20, 37, 48]. The most popular ones for study, in part because they are so common, are $NaAlH_4$ and $LiAlH_4$. As can be seen in **Table 1**, as HSMs, they can release H_2 via hydrolysis.

Varying degrees of success were achieved in efforts to demonstrate *reversible* thermal hydrogenation of $NaAlH_4$ and $LiAlH_4$, particularly when the materials were either nanoparticulates, or under the condition of nanoconfinement. First, it was demonstrated that the H_2 storage properties of these aluminum hydrides could be altered by microstructural refinement (particle and grain size) via the process of *ball milling* [49]. Furthermore, reversible hydrogenation of either $NaAlH_4$ or $LiAlH_4$ was demonstrated when: (1) Fullerene-C_{60} was used as a catalyst [50], (2) $NaAlH_4$ was confined within the nanoporous structure of titanium-loaded, highly-ordered, mesoporous carbon [51]. (3) Carbon nanotubes and graphitic nanofibers were used as catalysts [52]. (4) When a nanoparticulate-complex hydride mixture, stemming from the ad-mixing of $LiAlH_4$ and nano-MgH_2, was used as an MGE-HSM [53].

Outside of the methodology of using aluminum-containing compounds as MGE-HSMs, Omran and co-workers demonstrated the release of H_2 when aluminum chips[14] were exposed to aqueous base solutions [54].

---

[14] The advantage of this method is economic. One less processing step of terrestrial aluminum is required (see **Table 1**).

3.2.3 Silicon (cast form)

Terrestrial silicon consists of its oxide forms including silica (SiO_2) and the silicates. Silica is chemically transformed, on a production scale, into its native state, silicon metal, by its carbothermic reaction with charcoal at around 2500 deg. C [26–28]. This reaction is displayed in Eq. (4).

$$SiO_2 + 2\,C \rightarrow Si + CO + \tfrac{1}{2}\,CO_2 \tag{4}$$

Regarding the H_2 storage application, the chemical processes necessary to form Si-H bonds also need to be considered. In analogy to Eq. (4), silica will react with H_2 at high temperature (1400–1600 deg. C), and the silicon atom will go through a sequential reductive process, first yielding SiO, then Si, then compounds containing Si-H bonds. The by-product of these reactions is water, H_2O [40, 41].

3.2.4 Silicon (nanoparticulate)

Nanosilicon, due to its much larger surface area, is substantially more reactive than the cast form. If prepared correctly (yielding small enough particles and a surface free of oxidation), hydrogen is generated by the direct, heterogeneous reaction with water [55] as displayed in Eq. (5).

$$Si(nano) + 2\,H_2O \rightarrow SiO_2 + 2\,H_2 \tag{5}$$

Erogbogdo and co-workers [56] reported that the rate of H_2 generation from the hydrolysis of 10 nm silicon particles was 1000 times the rate when one used bulk (cast) silicon. A Si-containing material, specifically that of the "alkali metals plus silica gel" composites[15] developed by Dye and co-workers [57] releases H_2 gas when exposed to water under mild conditions. These composites have been commercialized as a H_2 fuel source and are marketed under the trade name "Active Fuel".

Methods demonstrated to produce water-reactive, H_2-generating, silicon-containing materials include: (1) The ball milling of mixtures; (a) The combination of silicon-containing precursors (such as $SiCl_4$) and various reducing agents (such as lithium metal) [58], (b) The combination of carbon (graphite) and SiO_2 [59] and (c) The combination of SiO_2 with aluminum [60], (2) The elevated temperature treatment of rice husk with aqueous sodium hydroxide such as the work of Bose et al. [61], and Liu et al. [62], and (3) The laser ablation of silicon wafers [63].

Finally, a process to produce commercial quantities of nanosilicon has been recently report by the firm PyroGenesis of Montreal, Canada [64]. PyroGenesis developed the process for their PUREVAP™ Nano Silicon reactor.

3.2.5 Silanes and polymers featuring a "backbone" containing silicon atoms

Si-H bond containing compounds including (1) Silanes, with the generic molecular structure SiX_aY_b and, specifically, the examples where the X group is a hydrogen atom, and (2) Polymers containing Si-H bonds, can be classified as MGE-HSMs. The polar, covalent, Si-H bond is hydridic and, when exposed to protonic "activators"

---

[15] Dye and his co-workers considered this material to be similar to nanosilicon.

under defined conditions, will generate H_2 gas. Generally, the rate of generation scales with the density of Si-H bonds in the material.

Monosilane, SiH_4, with the highest Si-H bond density of any of these types of materials, has long been considered a candidate MGE-HSM. Jackson and co-workers reported the synthesis of silane, SiH_4, from the reaction outlined in Eq. (6) [39].

$$3\,SiO_2 + 4\,Al + 2\,AlCl_3 + 6\,H_2 \rightarrow 6/n\,[AlOCl]_n + 3\,SiH_4 \quad (6)$$

A serious problem with utilizing SiH_4 as an MGE-HSM is that, like diborane (B_2H_6), silane is spontaneously explosive in air. Therefore, one method to use Si-H bond-containing materials as H_2-generating agents, while avoiding explosions, is to work with *polysilanes*, whose linear structure, analogous with linear hydrocarbons, has the general formula Si_xH_{2x+2}.[16] For example, as displayed in Eq. (7), Simone and co-workers demonstrated the thermal generation of H_2 from low MW polysilanes [65]:

$$Si_xH_{2x+2} \rightarrow Si\,(metal) + (x+1)\,H_2 \quad (7)$$

In related work, as displayed in Eq. (8), Brunel demonstrated the hydrolytic generation of H_2 from polysilanes [38]:

$$Si_4H_{10} + 12\,H_2O/NaOH \rightarrow \text{“}SiO_2\text{”} + 12\,H_2 \quad (8)$$

Linear polymers containing -SiH_2X- (X = CH_2, NH, O) monomeric groups could potentially function as MGE-HSMs. Polymers where X = O (“polysiloxanes” or “silicones”) were first reported by Rochow [66], Seyferth and co-workers [67, 68], Nishii and Narisawa [69], Harimoto and co-workers [70] and recently by Lin and co-workers [71]. As depicted in Eq. 10, conversion of -$(O\text{-}SiH_2)_x$- materials to silica (films), presumably with the release of hydrogen gas, occurred under a variety of conditions [70]:

$$-(O - SiH_2)_x - +2x\,H_2O \rightarrow -[O - Si(OH)_2]_x - +2\,H_2 \quad (9)$$

In a related study, Yap and co-workers reported the hydrolytic generation of H_2 from silicones (commercial products) containing the –(SiMeHO)- monomer [72].

Linear polymers containing -SiH_2X- where X = NH (“polysilazanes”) have been prepared and commercialized. The decomposition of select polysilazanes yield high quality[17] silica thin films, with presumably the generation of H_2 under mild conditions [73–76].

Linear polymers containing -SiH_2X- where X = CH_2 (“polycarbosilanes”) have been prepared [77]. Linear –$(H_2C\text{-}SiH_2)_x$- remains a liquid up to a substantial molecular weight. The heating of this liquid to approximately 300 deg. C results in the liberation of H_2 and the formation of a glassy solid. In a related study, when a stoichiometric amount of methanol was added to a mixture of cyclic organosilane, $(CH_2SiH_2)_3$ or $(CH_2SiH_2CHSiH_3)_2$, and 5 mol% NaOMe, rapid hydrogen release was observed at room temperature within 10–15 s. The original cyclic carbosilane could be regenerated via reaction with $LiAlH_4$ [78].

[16] To “be on the safe side”, one should work with polysilanes where X > 20 (authors’ opinion).

[17] The term “high quality” in this context refers to silica films that are both “dense” and “stable”.

4. Conclusions

A major challenge with using hydrogen, H_2, as an energy source, is finding methods by which to store it effectively and safely. In some applications, storing "pure" H_2, either as a compressed gas or a cryogenic liquid, is not feasible. This concern leads to the concept of employing storage materials which, via chemical reactions, release H_2 gas under mild, predictable and controlled conditions. Specifically, this review is concerned with main-group element-based hydrogen storage materials, MGE-HSMs, in either the native state of the element or else compounds featuring hydridic X-H bonds, where the central element (X) is one of the following seven; Lithium (Li), Sodium (Na), Potassium (K), Magnesium (Mg), Calcium (Ca), Boron (B), Aluminum (Al) and Silicon (Si).

High energy, often CO_2 generating, chemical processes are necessary to produce these MGE-HSMs from their terrestrial precursors, which are usually oxide, chloride, or silicate-based minerals. Once formed, the reactivity of the hydridic X-H bond to protonic reagents, usually water, a process that efficiently generates H_2 gas, scales with the degree of X^+H^- polarity of the bond. Regarding the Periodic Table of the Elements, this polarity increases either as one: (1) Descends a column of the table, or (2) Traverses from right-to-left across a row.

The Group I and II MGE-HSMs exist as the binary, ionic, hydride salts. The generation of H_2 from hydrolysis of these salts yields their common hydroxide compounds. The MGE-HSMs encompassing the hydrides of boron, aluminum and silicon are structurally complex and, in many cases, feature X-H bonds of significant covalent character which demonstrate reduced reactivity to protonic reagents. Furthermore, within the context of H_2 storage and generation; (1) Nanoscopic silicon metal, and a number of related composite materials, generate H_2 gas upon exposure to water at ambient temperatures, (2) Certain nanoscopic forms of $MAlH_4$ (M = Li, Na) demonstrate reversible, thermal hydrogenation/ dehydrogenation chemistries, and (3) Silicon-based polymers containing Si-H bonds release H_2 gas under specific hydrolytic/ thermolytic conditions.

Acknowledgements

The authors wish to thank Mr. Hans Haas, Mr. Kirk Ball and Dr. Sarah Snow for many useful discussions and insights.

Author details

Steven Snow*, Trisha Hoover and Malcolm Penman
AB Specialty Silicones, Waukegan, Illinois, USA

*Address all correspondence to: steve.snow@andisil.com

References

[1] Koonin S. Unsettled: What Climate Science Tells us, What it Doesn't and Why it Matters. Dallas, TX, USA: BenBella Books; 2021. ISBN 1-950-665-79-8

[2] Pahwa P, Pahwa G. Hydrogen Economy. New Delhi, India: The Energy Resources Institute (TERI) Press; 2014. ISBN: 9788179935040

[3] U.S. Department of Energy, Energy Information Agency. Annual Energy Outlook 2020 with Projections to 2050. 2018. Available from: heia.go/utlook/e/d/EO2020%20Full%20Report.pdf

[4] Dehganimadvar M, Shirmohammadi T, Sadeghzadeh M, Aslani A, Ghasempour R. Hydrogen production technologies: Attractiveness and future perspectives. International Journal of Energy Research. 2020;**44**: 8233-8254. DOI: 10.1002/er.5508

[5] Elgowainy A, Mintz M, Lee U, Stephens T, Sun P, Reddi K, et al. Energy systems division, US Department of Energy. Assessment of Potential Future Demands for Hydrogen in the United States. 2020. DOI: 10.2172/1710201

[6] Ruth M, Jadun P, Connelly E, Boardman R, Elgowainy A, Gilroy N, Simon A, Zuboy J. National Renewable Energy Laboratory; United States Department of Energy. The Technical and Economic Potential of the H2@Scale Concept within the United States. 2020. Available from: nrel.gov/docs/fy21osti/77610.pdf

[7] Züttel A. Hydrogen storage methods. Naturwissenschaften. 2004;**91**:157-172. DOI: 10.1007/s00114-004-0516-x

[8] Rodl & Partner. Ways of Financing Renewable Energy Projects in Emerging and Developing Countries. Newsletter of RE|nEws, 2016. Available from: roedl.com/insights/renewable-energy/2016-10/financing-renewable-energy-projects-emerging-developing-countries

[9] Kurtz J, Ainscough C, Simpson L, Caton M, U.S. Department of Energy. National Renewable Energy Laboratory (NREL). Hydrogen Storage Needs for Early Motive Fuel Cell Markets. 2012 NREL/TP-5600-52783. Available form: nrel.gov/docs/fy13osti/52783.pdf

[10] Ronnebro E, Stetson N, U.S. Department of Energy. Hydrogen and Fuel Cells Program. Early Market TRL/MRL Analysis Hydrogen Storage Technologies. 2013. PNNL-21580-HT020200. Available from: osti.gov/biblio/1108149

[11] Zhouri B. Hydrogen Energy, Challenges and Solutions for a Cleaner Future. Cham, Switzerland: Springer; 2016. DOI: 10.1007/978-3-319-93461-7

[12] Andersson J, Gr€onkvist S. Large scale storage of hydrogen. International Journal of Hydrogen Energy. 2019;**44**: 11901-11919. DOI: 10.1016/j.ijhydene.2019.03.063

[13] Lai Q, Yahui S, Wang T, Modl P, Cazorla C, Demirci U, et al. How to design hydrogen storage materials; fundamentals, synthesis and storage tanks. Advanced Sustainable Systems. 2019;**3**:1900043. DOI: 10.1002/adsu.201900043

[14] Guera O, Zhang J, Eichman J, Denholm P, Kurtz J, Hodge B-M. The value of seasonal energy storage technologies for the integration of wind and solar power. Energy and Environmental Science. 2020;**13**: 1909-1922. DOI: 10.1039/d0ee00771d

[15] Abdin Z, Tang C, Catchpol K, et al. Current state and challenges for hydrogen storage technology. In: Jaiswa-Nagar D, Dixit V, Devasahayam S, editors. Towards Hydrogen Infrastructure: Advances and Challenges in Preparing for the Hydrogen Economy. Amsterdam, Netherlands: Elsevier; 2023. DOI: 10.1016/C2021-0-03429-5

[16] Unknown author. Module 1, Hydrogen Properties, Revision 0. Course notes, College of the Desert 2001. Accessed 2024. Available from: https://www.energy.gov/sites/prod/files/2014/03/f12/fcm01r0.pdf

[17] Lewis P. Small containers-hydrogen explosions. In: Thai H-T, Calió I, Dong L, editors. Forensic Polymer Engineering: Why Polymer Products Fail in Service. Amsterdam, Netherlands: Elsevier, Woodhead Publishing in Materials; 2016. pp. 147-190. DOI: 10.1016/j.polymertesting.2010.03.002

[18] Cadwallader L, Herring J, U.S. Dept. of Energy. Idaho National Engineering and Environmental Laboratory. Safety Issues with Hydrogen as a Vehicle Fuel1999. Report # INEEL/EXT-99-00522. DOI: 10.2172/761801

[19] Hwang H, Varma A. Hydrogen storage for fuel cell vehicles. Current Opinion in Chemical Engineering. 2014; 5:42-48. DOI: 10.1016/j.coche.2014.04.004

[20] Prosini P. A Survey on Complex Hydrides for Hydrogen Storage. Saarbrucken, Germany: Lambert Academic Publishing; 2013. ISBN-13 978-3659373541

[21] Cotton F, Wilkinson G. Advanced Inorganic Chemistry. 5th ed. Hoboken, NJ, USA: Wiley Interscience; 1988. ISBN # 0 471-84997-9

[22] Patonia A, Poudineh R. Ammonia as a storage solution for future decarbonized energy systems. In: The Oxford Institute for Energy Studies (OIES) Paper EL 42. 2020. ISBN 978-1-78467-167-9

[23] Huheey J. Inorganic Chemistry: Principles of Structure and Reactivity. 3rd ed. Buffalo, NY, USA: Harper International; 1983. pp. 53-137. ISBN # 0-06-042987-9

[24] Olah G, Schlosberg R, Chemistry in Super Acids. I. Hydrogen exchange and polycondensation of methane and alkanes in $FSO_3H–SbF_5$ ("magic acid") solution. Protonation of alkanes and the intermediacy of CH_5^+ and related hydrocarbon ions. The high chemical reactivity of "Paraffins" in ionic solution reactions. Journal of the American Chemical Society. 1968;**90**:2726-2727. DOI: 10.1021/ja01012a066

[25] Unknown author. Petroleum. National Geographic. 2023. Available from: education.nationalgeographic.org/resource/petroleum/

[26] Brook M. Silicon in Organic, Organometallic and Polymer Chemistry. Hoboken, NJ, USA: Wiley-Interscience; 2000. ISBN # 0-471-19658-4

[27] Noll W. The Chemistry of Silicones. Cambridge, MA, USA: Academic Press; 1965. Library of Congress Card # 67-22772

[28] Eaborn C. Organosilicon Compounds. London, UK: Butterworths Scientific Publications; 1960

[29] Pahwa P, Pahwa G. Hydrogen Economy. New Delhi, India: The Energy Resources Institute (TERI) Press; 2014. pp. 137-141. ISBN: 9788179935040

[30] Cotton F, Wilkinson G. Advanced Inorganic Chemistry. 5th ed. Hoboken, NJ, USA: Wiley Interscience; 1988. pp. 118-125. ISBN # 0 471-84997-9

[31] Cotton F, Wilkinson G. Advanced Inorganic Chemistry. 5th ed. Hoboken, NJ, USA: Wiley Interscience; 1988. pp. 168-178. ISBN # 0 471-84997-9

[32] Prosini P. A Survey on Complex Hydrides for Hydrogen Storage. Saarbrucken, Germany: Lambert Academic Publishing; 2013. pp. 161-166. ISBN-13 978-3659373541

[33] Beaird AM. Mechanisms in the Water Vapor Hydrolysis of Sodium Borohydride. Columbia, SC, USA: University of South Carolina; 2011

[34] Cotton F, Wilkinson G. Advanced Inorganic Chemistry. 5th ed. Hoboken, NJ, USA: Wiley Interscience; 1988. pp. 208-221. ISBN # 0 471-84997-9

[35] Prosini P. A Survey on Complex Hydrides for Hydrogen Storage. Saarbrucken, Germany: Lambert Academic Publishing; 2013. pp. 75-88. ISBN-13 978-3659373541

[36] Holleman A, Wiberg F, Wiberg N. Lehrbuch der Anorganischen Chemie. Vol. 102. Berlin, Germany: de Gruyter; 2007. DOI: 10.1515/9783111520285

[37] Ashton E, Oakley W, Brack P, Dann S. Evaluation of the vapor hydrolysis of lithium aluminum hydride for Mobile fuel cell applications. ACS Applied Energy Materials. 2022;**5**: 8336-8345. DOI: 10.1021/acsaem.2c00891

[38] Brunel J. Polysilanes: The grail for a highly-neglected hydrogen storage source. International Journal of Hydrogen Energy. 2017;**42**: 23004-23009. DOI: 10.1016/j.ijhydene.2017.07.206

[39] Jackson H, Marsh F, Muetterties E. Chemistry of Silane. I. Preparation of Silane. Inorganic Chemistry. 1963;**2**: 43-46. DOI: 10.1021/ic50005a013

[40] Tso S, Pask J. Reaction of fused silica with hydrogen gas. Journal of the American Ceramic Society. 1982;**65**: 457-460. DOI: 10.1111/j.1151-2916.1982.tb10514.x

[41] Accolla M et al. Silicon and hydrogen Chemistry under laboratory conditions mimicking the atmosphere of evolved stars. The Astrophysical Journal. 2021; **906**(1):44-50. DOI: 10.3847/1538-4357/abc703

[42] Author U. Abundance of the Earth's elements in the Earth's crust and in the sea. In: Lide D, editor. Boca Raton, FL, USA: CRC Handbook of Chemistry and Physics. 80th ed. 1999-2000. p. 14-14. ISBN # 0-8493-0480-6

[43] Stock A. Hydrides of Boron and Silicon. Ithaca, NY: Cornell University Press; 1957. ASIN B000856QUU

[44] Lipscomb W. Boron Hydrides. Mineola, NY, USA: Dover Books; 2012. ISBN-13 978-0-486-48822-6

[45] Demirci U. Ammonia borane: An extensively studied, though not yet implemented, hydrogen carrier. Energies (mdpi). 2020;**13**:1-45. Available from: www.mdpi.com/journal/energies

[46] Severa G, Ronnebro E, Jensen C. Direct hydrogenation of magnesium boride to magnesium borohydride: Demonstration of >11 weight percent reversible hydrogen storage. Chemical Communications. 2010;**46**:421-423. DOI: 10.1039/B921205A

[47] Ouellette R, Rawn J. Organic Chemistry; Structure, Mechanism and Synthesis. 2nd ed. Amsterdam, Netherlands: Elsevier; 2019. DOI: 10.1016/C2016-0-04004-4

[48] Graetza J, Reilly J, Yartys V, Maehlen J, Bulychevc B, Antonov V, et al. Aluminum hydride as a hydrogen and energy storage material: Past, present and future. Journal of Alloys and Compounds. 2011;**509S**:S517-S528. DOI: 10.1016/j.jallcom.2010.11.115

[49] Varin R, Zbroniec L, Polanski M, Bystrzycki M. A review of recent advances on the effects of microstructural refinement and Nanocatalytic additives on the hydrogen storage properties of metal and complex hydrides. Energies. 2011;**4**(1):1-25. DOI: 10.3390/en4010001

[50] Terpovich J, Knight D, Wellons M, Zidan R. Catalytic effect of fullerene and formation of nanocomposites with complex hydrides: $NaAlH_4$ and $LiAlH_4$. Journal of Alloys and Compounds. 2011; **509**:S562-S566. DOI: 10.1016/j.jallcom.2010.10.031

[51] Xiong R, Sang G, Yan X, Zhang G, Ye X. Improvement of the hydrogen storage kinetics of $NaAlH_4$ with Ti-loaded high-ordered mesoporous carbons (Ti-OMCs) by melt infiltration. Journal of Materials Chemistry. 2012;**22**: 17183-17189. DOI: 10.1039/C2JM31961F

[52] Hudson M, Raghubanshi H, Pukazhselvan D, Srivastava O. Carbon nanostructures as catalyst for improving the hydrogen storage behavior of sodium aluminum hydride. International Journal of Hydrogen Energy. 2012;**37**:2750-2755. DOI: 10.1016/j.ijhydene.2011.03.006

[53] Vittetoe A, Niemann M, Srinvasan S, McGrath K, Kumar A, Goswami D, et al. Destabilization of $LiAlH_4$ by nanocrystalline MgH_2. International Journal of Hydrogen Energy. 2009;**34**: 2333-2339. DOI: 10.1016/j.ijhydene.2009.01.025

[54] Omram M, Oskueyan G, Sedighi K, Farhadi M. Design of Hydrogen Generator for on-board hydrogen generation from waste aluminum chips. In: Proceedings of the 22nd International Symposium on Transport Phenomena. 2011

[55] Goller B, Kovalev D, Sreseli O. Nanosilicon in water as a source of hydrogen: Size and pH matter. Nanotechnology. 2011;**22**:305402. DOI: 10.1088/0957-4484/22/30/305402

[56] Erogbogbo F, Lin T, Tucciarone O, LaJoie K, Lai L, Patki G, et al. On-demand hydrogen generation using nanosilicon: Splitting water without light, heat or electricity. ACS Nano. 2013;**13**:451-456. DOI: 10.1021/nl304680w

[57] Dye J, Cram K, Urbin S, Redko M, Jackson J, Lefenfeld M. Alkali metals plus silica gel: Powerful reducing agents and convenient hydrogen sources. Journal of the Chemical Society, Chemical Communications. 2005;**127**: 9338-9339. DOI: 10.1021/ja051786+

[58] Chaudhary A, Sheppard D, Paskevicius M, Saunders M, Buckley C. Mechanochemical synthesis of amorphous silicon nanoparticles. RSC Advances. 2014;**4**:21979-21983. DOI: 10.1039/C3RA47431C

[59] Lam C, Zhang Y, Tang Y, Lee C, Bello I, S. Lee S. Large-scale synthesis of ultrafine Si nanoparticles by ball milling. Journal of Crystal Growth. 2000;**220**: 466-470. DOI: 10.1016/S0022-0248(00)00882-4

[60] Araujo-Andrade C, Espinosa-Beltran F, Jimenez-Sandoval S, Gonzalez HJ. Synthesis of nanocrystalline Si particles from a solid-state reaction during a ball-milling process. Scripta Materialia. 2003; **49**:773-778. DOI: 10.1016/S1359-6462(03)00429-9

[61] Bose S, Ganayee M, Mondal B, Baidya A, Chennu S, Mohanty J, et al. Synthesis of silicon Nanoparticles from Rice husk and their use as sustainable fluorophores for white light emission. ACS Sustainable Chemistry and Engineering. 2018;**6**:6203-6210. DOI: 10.1021/acssuschemeng.7b04911

[62] Liu N, Huo K, McDowell M, Zhao J, Cui Y. Rice husks as a sustainable source of nanostructured silicon for high performance Li-ion battery anodes. Scientific Reports. 2013;**3**:1919. DOI: 10.1038/srep01919

[63] Huan C, Shu-Qing S. Silicon nanoparticles: Preparation, properties and applications. Chinese Phys. B. 2014; **23**:088102-(1-14). DOI: 10.1088/1674-1056/23/8/088102

[64] GlobeNewswire. Pyrogenesis Confirms Successful Production of Nano Silicon Material. GlobeNewswire.com. 2021. Available from: globenewswire.com/news-release/2021/08/05/2275933/0/en/PyroGenesis-Confirms-Successful-Production-of-Nano-Silicon-Material.html [Accessed: February 12, 2023]

[65] Simone D, Claudio B, Hidding B. Silanes as Fuel for Aerospace Propulsion. Transactions of Space Technology (Japan). 2009;7:33-39. DOI: 10.2322/tstj.7.Pa_33

[66] Rochow E. An Introduction to the Chemistry of the Silicones. Hoboken, NJ, USA: Wiley; 1946

[67] Seyferth D, Prud'homme C, Wiseman G. Cyclic Polysiloxanes from the hydrolysis of Dichlorosilane. Inorganic Chemistry. 1983;**22**:2163-2167. DOI: 10.1021/ic00157a014

[68] Seyferth D, Prud'homme C. Linear Polysiloxanes from Dichlorosilane. Inorganic Chemistry. 1984;**23**: 4412-4417. DOI: 10.1021/ic00194a003

[69] Nishii K, Narisawa T. Production of polydihydrogensiloxane. Japan Patent . JPS6042426A. 1983

[70] Harimoto Y, Kushibiki N, Itoh M, Katsoulis D. Cyclic dihydrogenpolysiloxanes, hydrogenpolysiloxanes, processes for their production, silica type glass moldings and a process for their production, optical elements and a process for their production. United States of America Patent US8097690B2. 2012

[71] Lin Y et al. Characterization of shock-sensitive deposits from the hydrolysis of Hexachlorodisilane. ACS Omega. 2019;**4**:1416-1424. DOI: 10.1021/acsomega.8b03103

[72] Yap C, Poh H, Fan W. Metal-free catalytic hydrogen production from a polymethylhydrosilane–water mixture. RSC Advances. 2016;**6**:5903-5590. DOI: 10.1039/C5RA23887K

[73] Lee J-S, J-H OH, Moon S-W, Sul W-S, Kim S-D. A technique for converting Perhydropolysilazane to SiOx at low temperatures. Electrochemical and Solid State Letters (The Electrochemical Society). 2010;**13**:H23-H25. DOI: 10.1149/1.3264092

[74] Smokovych I, Gatzen C, Kruger M, Schwidder M, Scheffler M. Polymer derived ceramics from Si, B, SiB6, and Mo5SiB2 filler-loaded

Perhydropolysilazane precursors as protective and functional coatings for refractory metal alloys. Materials. 2020; **2020**(13):4878. DOI: 10.3390/ma13214878

[75] Zongbo Z, Shao Z, Luo Y, An P, Zhang M. Hydrophobic, transparent and hard oxynitride coating from perhydropolysilazane. Polymer International. 2015;**2015**(64):971-978. DOI: 10.1002/pi.4871

[76] Isoda T, Kaya H, Nishii H, Funayama O, Suzuki T, Tashiro Y. Perhydrosilazane precursors to silicon nitride ceramics. Journal of Inorganic and Organometallic Polymers. 1992; **1992**(2):151-160. DOI: 10.1007/BF00696542

[77] Interrante L, Shen Q. Polycarbosilanes. In: Jones R, Ando W, Chojnowski J, editors. Silicon-Containing Polymers: The Science and Technology of Their Synthesis and Applications. Dordrecht, Netherlands: Kluwer Academic Publishers; 2000. pp. 247-321. ISBN # 9780412831102

[78] Han W-S, Kim T-J, Kim S-K, Kim Y, Kim Y, Nam S-W, et al. Silane-based hydrogen storage materials for fuel cell application: Hydrogen release via methanolysis and regeneration by hydride reduction from organosilanes. International Journal of Hydrogen Energy. 2011, 2011;**36**:12305. DOI: 10.1016/j.ijhydene.2011.06.118

Section 4

Hydrogen Transportation

Chapter 5

Transportation of Hydrogen: Hydrogen Usage

Akbar Dauletbay

Abstract

For large-scale hydrogen use for alternative fuel problems, hydrogen transportation must be solved. Hydrogen can be transported as compressed gas, liquid, or bound in carriers. The chapter describes current transportation technologies—gaseous hydrogen *via* pipelines or special trucks, and liquid hydrogen in cryogenic tanks. The potential of using existing natural gas pipelines is analyzed; the need for modern pipeline material complex research is emphasized. Transportation in solid or liquid carriers, disadvantages and advantages of transportation methods, and problems and ways to solve them are analyzed. Hydrogen facilitates the conversion of low-grade crude oils into high-energy transport fuels by catalytic cracking and desulfurization. Ammonia production, essential for fertilizers and explosives, relies heavily on hydrogen synthesis from nitrogen and hydrogen. Methanol and dimethyl ether fuels offer alternatives to hydrogen storage and transportation, while liquid hydrocarbon fuels from coal and biomass utilize hydrogen in conversion processes like Fischer-Tropsch. Proton exchange membrane and alkaline fuel cells depend on hydrogen for electricity generation in transportation. Additionally, hydrogen serves as a reductant in metallurgy, with advancements in direct iron reduction and green steel initiatives driving sustainable practices in the steel industry. These applications underscore in modern processes and its potential for addressing energy and environmental challenges.

Keywords: transportation technologies, pipeline network, carriers, liquid hydrogen, usage of hydrogen, fuel cells

1. Introduction

The environment has suffered from the greenhouse gas emissions (CO_2, NO_2, CH_4, and O_3) that result from burning natural resources (natural gas, coal, and oil), which cause global warming. Moreover, these natural resources will soon run out due to their finite availability in nature. Hence, a new way of producing energy that is clean and sustainable is needed to ensure future energy security, which requires a gradual shift away from natural resources. To address these challenges, various renewable energy sources such as wind, solar, and nuclear have been investigated recently. However, these sources cannot meet the global energy demand due to weather and location limitations. In addition, this energy is not always accessible and requires transportation. Therefore, hydrogen has been the focus of extensive research as an energy carrier. Also, this energy is not always available and needs to be solved in terms of transportation as well as production and storage issues.

Hydrogen is the most plentiful element in the universe; it has high energy efficiency and is eco-friendly [1, 2]. Hydrogen is the energy carrier [2–4], which means it can store and deliver electrical energy through chemical reactions instead of combustion [5]. It can also be easily implemented in transportation to fuel cars, heat homes, and many other applications [6]. Its byproduct is only water and heat. Hydrogen has a higher energy content per mass (120 MJ/kg) than 44 MJ/kg for gasoline [7, 8]. However, hydrogen needs to be produced in a cost-effective way before it can be used in a practical form for application. Hydrogen is not naturally available in nature. Therefore, it needs to use a primary energy source, such as non-renewable energy sources like fossil fuel and renewable energy sources like solar energy, wind energy, and biomass. Also, hydrogen can be used instead of secondary energy sources like electricity or heat. In addition to the production and storage issues, the hydrogen that is produced must be transported safely for future use. Therefore, various features need to be considered, such as its high flammability limit in the air (4–74%) [9] compared to gasoline vapor (1.4–7.6%) and natural gas (5.3–15%). Also, its high explosion limit in the air (H_2 = 18.3–59%) compared to gasoline vapor (1.1–3.3%) and natural gas (5.7–14%). In addition, its low ignition energy (0.02 MJ) [9] compared to gasoline vapor (0.20 MJ) and natural gas (0.29 MJ) has to be controlled. Finally, its lower boiling point (−253°C) and low density in the liquid state (70.8 g/L) [10] compared to gasoline vapor (37e205°C) with a density of 700 g/L and natural gas −162°C) with a density of 423 g/L, require additional safety measures for hydrogen fuel.

Hydrogen is a versatile and clean energy carrier that can be used for various applications. It is the simplest and most abundant element in the universe, but it is not naturally available in a pure form on Earth. Hydrogen can be stored and transported in different ways, such as in pressurized or cryogenic tanks, pipelines, or blended with natural gas. Hydrogen can also be converted into electricity or heat through fuel cells or combustion or used as a chemical feedstock for many industries. Some of the benefits of hydrogen are that it has a high energy density, it produces no harmful emissions when used, and it can be integrated with renewable energy sources. However, some of the challenges of hydrogen are that it requires high capital costs, it has safety issues due to its flammability and low density, and it has a low efficiency of conversion and storage [11–14]. Therefore, hydrogen is a promising energy carrier that needs further research and development to overcome its technical and economic barriers. According to the safety guideline, hydrogen can be transported traditionally in gaseous or liquid form in the pressurized or cryogenic tank. Also, finding a suitable means to transport hydrogen is a crucial part of the hydrogen economy. There are various means of hydrogen transport such as pipeline, blending natural gas, and cryogenic liquid tankers. Therefore, this chapter will try to summarize the state of the art of primary hydrogen transportation methods and uses of hydrogen. Also, identify the most promising techniques to improve hydrogen usage and transportation developments.

2. Transportation of hydrogen

Determining an effective mode of hydrogen transportation is a crucial step for implementing hydrogen fuel across diverse industrial applications. The transportation of hydrogen requires a thorough understanding of its properties to prevent potential explosion and leakage incidents. Given that hydrogen can exist in different states based on temperature and pressure, the transportation methods vary.

At low temperatures, hydrogen assumes a solid state with a density of 70.6 kg/m^3 at −262°C, while at higher temperatures, it transforms into a gas with a density of 0.089 kg/m^3 at 0°C and 1 bar. In the region highlighted in blue in **Figure 1**, between the triple point (13.8 K) and the critical point (33 K), hydrogen exists in a liquid state with a density of 70.8 kg/m^3 [15]. At standard temperature (25°C) and 1-atm pressure, hydrogen is a gas, and its low critical temperature (T_c = 33 K) is attributed to the strong repulsion interaction between H_2 molecules. Consequently, storing and transporting H_2 necessitate addressing the challenge of the large volume occupied by hydrogen gas. For instance, 1 kg of H_2 at standard temperature and atmospheric pressure fills a volume of 11 m^3. While the Joule-Thompson effect causes a pressure reduction in natural gas leading to a temperature drop of 0.5°C, hydrogen experiences an increase in temperature of 0.35°C for each rise in bar pressure. Therefore, the development of a hydrogen transportation system requires careful consideration of various conditions to ensure safety during transportation. Various methods for hydrogen transportation have been proposed, with the most common ones being compressed gas cylinders, cryogenic liquid tankers, pipelines, and blending with natural gas.

Transporting hydrogen to its intended destination poses challenges due to its status as the least dense gas and its flammability when combined with even small amounts of air. The low volumetric energy density of hydrogen makes its transportation, storage, and eventual delivery to the point of use potentially costly. Safety concerns also arise in this process. Currently, hydrogen is predominantly conveyed

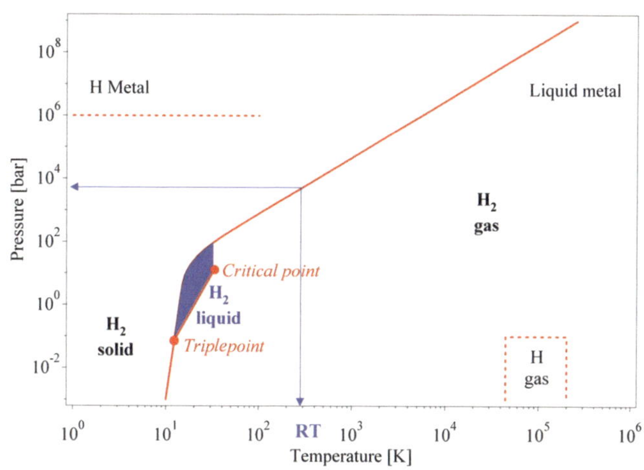

Figure 1.
Various states of hydrogen under different temperature and pressure conditions [15].

from its production site to the utilization site either through pipelines or over the road, utilizing liquid tanker trucks or gaseous tube trailers and special carriers.

2.1 Pipelines

It is worth noting that throughout the operational history of hydrogen pipeline systems in North America and Europe, no issues related to hydrogen embrittlement or safety have been reported [16]. However, the considerable capital investments required for pipeline construction make this method the most costly and feasible only for consistent and significant hydrogen consumption scenarios, where the pipeline construction expenses can be recouped within an acceptable timeframe.

The recommended pressure for the primary transportation of hydrogen, considering its physicochemical properties, is in the range of 7–14 MPa [17]. For instance, in the USA, hydrogen pipelines operate within the range of 3.5–10 MPa [18]. Distribution networks, with smaller pipe diameters, operate under lower pressure conditions (in the USA, p = 0.03 to 1.4 MPa [18]). However, gas stations and power plants require higher inlet pressures, suggesting that the pressure in distribution networks should be higher than in natural gas distribution lines, falling within the range of 1.4–2.8 MPa [18].

In low-pressure pipelines (0.1 MPa and below), the gas speed is 10 m/s, while in main pipelines (6 to 8 MPa), it is twice as high [17, 19]. With identical pipe diameters and pressure drop, the flow rate of hydrogen is nearly three times higher than that of methane. The specific cost of hydrogen transportation decreases with an increase in distance. For example, when the distance increases from 8 to 100 km, the cost decreases by an order of magnitude.

The construction costs for new hydrogen gas pipelines are relatively high, with labor and material expenses constituting around 70% of the total construction costs. Consequently, current priorities include developing new metallic, non-metallic, and composite materials, along with advancing technologies for applying thin barrier coatings to pipe surfaces.

The internal coating is designed to diminish the surface concentration of hydrogen on steel. Research on hydrogen diffusion in a multilayer pipe, featuring an internal coating based on reinforced polyamide, and external coatings made of polyurethane, has indicated that existing polymer and fiberglass materials may not extend the service life of pipelines by more than 10 years.

Reinforced plastic pipelines present a promising alternative to steel pipelines in terms of technical characteristics and cost. Typically, they consist of (1) an internal impermeable barrier pipe or liner, (2) a protective coating, (3) an intermediate coating, (4) composite layers made of glass or carbon fibers, (5) an external barrier layer, and (6) a protective coating (**Figure 2**). These pipes exhibit high compressive strength, can endure longitudinal deformations, facilitating their transport, and can be wound on large diameter reels (**Figure 3**). The multilayer design can incorporate sensors for real-time condition monitoring.

Polymer materials like polyethylene, polyamide, and polyvinylidene difluoride can be used for liners, and the hydrogen permeability of these materials determines the potential hydrogen leakage from the pipeline. Although most tests are conducted on films, the results may not universally apply to actual liners. Comparisons of permeability measurements for high-density polyethylene samples used in pipes and liners with published data for films indicate that the hydrogen losses from such pipelines are expected to be minimal—less than 0.1% of the transmitted volume [20]. The total

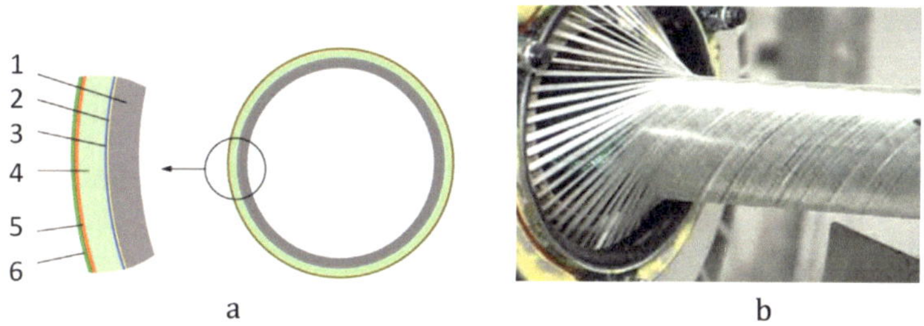

Figure 2.
Diagram of a multilayer plastic pipe (a) Fiberspar LinePipe, LLC and the fiber winding stage (b) [20].

Figure 3.
Fiberspar LinePipe [21].

capital investment for a polyethylene pipeline of this nature is approximately equivalent to that of a pipeline made of 16-inch steel pipes [22].

The pipeline system serves as a transportation network for natural gas or oil, connected by compressor stations, city gate stations, and storage facilities. Particularly suitable for large power plants (around 1000 metric tons/day), pipelines offer a cost-effective option [23]. Compressor stations utilize the heat from the transmission system to maintain consistent gas flow rates and pressures, meeting the specified requirements. The pipeline network encompasses both onshore and offshore components, covering transmission, transportation, and distribution pipelines. Utilizing pipelines for hydrogen transport is viewed as the most efficient means for extensive delivery and utilization as an energy carrier, presenting various advantages.

Hydrogen transportation through pipelines proves to be the most cost-effective for large-scale power plants, with a savings of $2.73 per kg of hydrogen transport [24]. Additionally, large-scale pipeline transport is considered the most environmentally friendly method of hydrogen delivery [24]. The pipeline's longevity, spanning several decades, contributes to its reputation for safety and reliability, given that most pipelines are buried underground. This minimizes the likelihood of accidents due to leakage, explosions, or environmental interference, avoiding disruptions and traffic on roads. Despite the significant initial capital investment for pipeline installation, subsequent maintenance and operation costs are comparatively low. Moreover, existing pipelines can transport pure hydrogen, and new pipelines for hydrogen delivery can be manufactured using low carbon steel.

However, challenges persist in establishing a hydrogen pipeline infrastructure. Hydrogen gas in a pipeline may experience losses compared to other fuels, and the need for compressing hydrogen to high pressures (around 10–20 bars) to enhance delivery speed poses a logistical hurdle due to its low density (1/8th of natural gas). The porosity of polymer materials used in gas pipelines makes them unsuitable for delivering pressurized hydrogen, as hydrogen's efficient escape, given its small size, poses safety risks. Furthermore, the embrittlement of pipeline steels and construction materials, leading to degradation and cracking, poses a risk of pipeline failure, dependent on material and operating conditions [25]. To address these issues, an alternative approach involving blending hydrogen with natural gas is proposed to mitigate risks and ensure a more secure distribution and delivery of hydrogen.

The first main hydrogen pipeline was put into operation in 1938 in Germany. This pipeline has been in operation for more than half a century without any accidents [19]. As of March 2023, Germany possessed the most extensive planned hydrogen transmission pipeline network in Europe, spanning a total of 3827 km [26]. Based on available information, the construction of hydrogen pipelines is gaining momentum, propelled by the increase in green hydrogen initiatives in China. A collective 1000 km of hydrogen pipelines are currently in the construction phase. This includes the development of two long-distance pipelines along with several shorter-distance pipelines [27].

Like natural gas currently, gaseous hydrogen has the capability to be transported through pipelines. The SoutH2 Corridor is a project that aims to connect to a "European Hydrogen Backbone" that will help Europe to achieve its green energy goals. The Backbone plan expects Europe to have 11,600 km of hydrogen pipelines by 2030 and almost 40,000 km by 2040 (**Figure 4**) [28].

China is currently constructing all of Asia's hydrogen pipelines, with three lines in the pipeline—Ulanqab Beijing, Shandong Hydrogen, and Ningxia Hui Autonomous Region Hydrogen. Additionally, there are plans for a hydrogen pipeline in the proposed India-Middle East-Europe Economic Corridor, targeting exports to the EU.

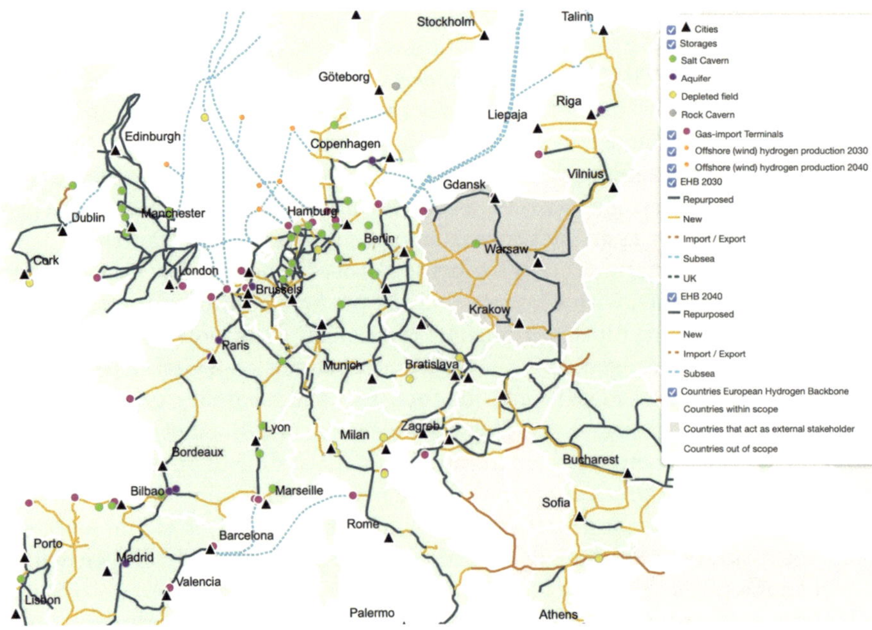

Figure 4.
Map of Europe's hydrogen pipeline plan until 2040 [28].

Over the past year, the Middle East has significantly increased its planned clean hydrogen capacity, the region has 83 low carbon or renewable hydrogen/ammonia projects with combined production of nine million metric tons hydrogen per year, S&P Global Commodity Insights' data show, and aims to become a key exporter to other countries by 2030. Saudi Arabia aims to be the leading global hydrogen supplier, while the UAE and Qatar are actively enhancing their capacities to meet the growing global demand [28].

In North America, both the United States and Canada are considering the implementation of hydrogen pipelines. The U.S. currently possesses 2600 km of hydrogen pipelines, and its HyBlend initiative is investigating methods to repurpose existing natural gas pipelines for hydrogen transportation. These pipelines are in places where there are a lot of hydrogen users, such as oil refineries and chemical factories, especially in the Gulf Coast area.

Canada's hydrogen initiatives involve a project in Quebec, anticipated to contribute to a 3% reduction in the province's carbon emissions over the current decade.

Latin America's significant potential for renewable energy positions it as a prominent provider of cost-effective, environmentally friendly hydrogen, according to the International Energy Agency (IEA). In the region, a total of 11 countries have devised strategies for hydrogen.

Chile aims to achieve the production of the world's most economical hydrogen by 2030 and strives to be among the top three hydrogen exporters by 2040. Similar to the United States and numerous other nations, Chile is exploring the possibilities of utilizing its natural gas pipelines for the safe transportation of hydrogen or for blending hydrogen with natural gas [28].

The World Economic Forum's Industrial Clusters initiative is establishing global hubs for hydrogen-related activities, fostering collaboration among stakeholders throughout the entire hydrogen value chain and uniting them around shared objectives. This initiative facilitates cluster members in accessing hydrogen from various producers, providing suppliers with a readily available pool of potential customers.

2.2 Compressed gas containers

Hydrogen gas is usually transported in cylindrical steel containers under pressure up to 20 MPa [19]. Such containers are delivered to the place of hydrogen consumption on automobile or railway platforms. Canadian company FIBA Canning Inc. offers various trailers with cylinders that can transport approximately 100 to 700 kg of hydrogen at pressures of 16–24 MPa (**Figure 5**) [29]. The cost of transporting compressed hydrogen by truck is quite high—slightly less than by pipeline, due to the low density of hydrogen.

Trailers for transporting hydrogen under pressure are effective in meeting the needs of small consumers, and the high cost of delivery can be offset by the absence of losses [30]. Currently, delivering hydrogen gas by trailer is the easiest way, especially in areas where there are no pipelines [30]. It is also convenient for delivery to fueling stations, where hydrogen trailers remain on site, without the need for permanent hydrogen gas storage infrastructure.

Recently, some researchers are also considering the option of delivering "cold gas" by trucks (trailers). For example, it is proposed to transport hydrogen gas at 35 MPa and 90 K in composite container pipes on trailers [31]. This will increase capacity and at the same time reduce liquefaction costs. The method is promising for delivering hydrogen to gas stations.

Figure 5.
Transportation of compressed hydrogen gas [29].

Depending on the required quantity, truck transportation is a feasible method for moving gaseous hydrogen in moderate amounts using compressed gas containers, such as cylinders or tubes pressurized within the range of 200–500 bar. In assessing the viability of hydrogen transportation by truck, factors such as transport capacity, tank weight, greenhouse gas emissions, and non-renewable energy consumption need careful consideration. For larger quantities, multiple pressurized gas cylinders or tubes are typically mounted on specialized trailers known as compressed gas hydrogen tube trailers, securely enclosed within protective frames for safety. The maximum transportable hydrogen load depends on the high weight of these cylinders or tubes. For example, a tube trailer equipped with steel cylinders can store up to 25,000 liters of hydrogen compressed to 200 bar, equivalent to 420 kg of H_2 [29].

To enhance the transported hydrogen quantity, lighter tank materials, such as composite materials for gas cylinders or tubes, are designed to handle higher pressures, allowing for the transportation of larger quantities of hydrogen per trailer. The cost-effectiveness of transporting hydrogen with tube trailers without liquefaction is evident, with a savings of $2.86 per kg delivered H_2 in small-scale power plants [30]. For instance, superlight cylinder materials made of carbon fiber composite with high-density polyethylene liners can accommodate up to 39,600 liters of hydrogen. These containers, pressurized to a maximum of 200 bar, can carry about 666 kg of H_2 [31].

In the case of transporting liquid organic hydrogen carriers (LOHC), trucks wait during the unloading and loading process, requiring only one trailer per truck. However, storage tanks are necessary at the hydrogenation and dehydrogenation sites for the liquid organic hydrogen carriers. Additionally, the LOHC delivery chain is reported to significantly enhance the economics of long-distance road transport [31].

2.3 Cryogenic liquid tankers

Various alternatives, including the transport of liquid hydrogen, have been suggested to address the challenges associated with compressed gas containers. Hydrogen, in liquid form, can be conveyed using trucks or other transportation modes. In contrast to compressed gas container, a liquid hydrogen trailer has the advantage of carrying more hydrogen due to the higher density of liquid hydrogen compared to hydrogen in a gaseous state. However, before storage in large insulated tanks at the liquefaction plant, gaseous hydrogen undergoes liquefaction by cooling it below −253°C through a process known as the liquefaction process [32]. The truck responsible for transporting liquid hydrogen is referred to as a liquid tanker. The effectiveness of thermal insulation significantly impacts the tank's operational parameters and operating costs [33].

Nevertheless, the road shipment of liquid hydrogen and its dispensing at a vehicle filling site add costs ranging from \$2.42 to \$1.40 per kg of H_2 to the production costs. Additionally, approximately 40% of energy is lost during the liquefaction process [34]. Furthermore, there is a loss of stored hydrogen through evaporation or boil-off of liquefied hydrogen, which is more pronounced when using a small tank with large surface-to-volume ratios. Lowering the liquefaction cost for hydrogen can have a positive impact on its shipment cost via truck, ship, or rail, and can also be advantageous for storage at plant sites to prevent plant shutdowns. This approach holds promise for near-term investments, particularly as the shipment plan can be a potential investment in the early stages of fuel cell vehicle introduction. The hydrogen boil-off point is a critical consideration during its delivery [35].

The creation of cryogenic complexes for hydrogen liquefaction, its long-term storage and transportation by railways and highways began in the 1960s of the last century in connection with the use of liquid hydrogen as fuel for rocket and space systems [36].

Hydrogen liquefaction is a very energy-intensive process and, therefore, expensive, but transportation costs for liquid hydrogen are minimal. The technology of hydrogen transportation by road, including safety measures, has been sufficiently developed. In the USSR, tank trucks TRZHV-20 (capacity 20 m^3) and TRZHV-24 (24 m^3) were created to transport liquid hydrogen over long distances [17]. Currently, JSC "Cryogenmash" produces custom-made tank cars with a capacity of 25 and 45 m^3 for the transportation of liquid hydrogen (**Figure 6**) [36, 37].

Figure 6.
Automobile tank of JSC "Cryogenmash" for liquid hydrogen transportation [37].

The tank is equipped with sophisticated systems for refueling and dispensing liquid hydrogen that meet Russian and European safety requirements and include a set of safety valves, rupture membranes, purge lines with pure nitrogen gas or vacuuming before refueling with hydrogen. In addition, the tank is equipped with effective wave dampers.

When transporting liquid hydrogen in tank trucks, losses caused by continuous evaporation of hydrogen and due to the performance of technological operations are inevitable. During one-time cooling of a tanker truck, up to 15% of hydrogen is lost from the volume of the tank, and cooling is carried out at least two times a year. Losses due to imperfections in the vacuum thermal insulation of the tank are 0.5%/day from its volume. Taking into account the fact that not all hydrogen is taken from the tank (a certain amount of liquid hydrogen remains for cooling), then for a tank with a capacity of 4.5 tons, the losses are about 8.2 tons/year.

At each refueling of the tanker, there are losses associated with the evaporation of the first portion of hydrogen. According to estimates, this is ~4%, that is, with a weight of 4.5 tons, they amount to ~180 kg. The losses for creating a pressure drop between the liquefaction plant and the capacity are approximately 1.5% [17].

Tanks for liquid hydrogen are made either cylindrical or spherical. Large containers are usually made spherical to reduce evaporation losses.

Liquid hydrogen is transported by tankers to a distance of more than 1.6 thousand km.

BMW specialists have created several prototype models of cars powered by liquid hydrogen fuel stored in special cylinders, in which the loss of hydrogen mass by evaporation is reduced to 1.5%/day [38]. BMW considers liquefied petroleum gas to be the most convenient type of fuel for promising cars.

Railway transport for the transportation of liquid hydrogen is used rather sparsely. In refrigerated railway tanks, hydrogen losses are about the same as in tank trucks. At present, JSC Cryogenmash offers customers high-speed hydrogen tanks with a capacity of 100 m^3. The tanks are equipped with a reinforcement cabinet and devices that ensure the safety of transportation [37].

Kawasaki Heavy Industries in Japan has developed a ship capable of holding 160,000 m^3 of liquefied hydrogen, equal to 11,200 tons, making it similar in size to a typical liquefied natural gas (LNG) carrier. This vessel would be 128 times larger than the tank on Kawasaki's Suiso Frontier, which transported the world's first liquefied hydrogen cargo from Australia to Japan in February.

Prior to this, only within the framework of the NASA space program, liquid hydrogen for refueling launch vehicles was transported on a special barge at a distance of about 100 km. However, the USA, Japan, South Korea and other countries have extensive experience in transporting liquefied natural gas in tankers. This experience will certainly be used in the creation of marine tankers for the transportation of liquid hydrogen.

2.4 Blending with natural gas

The incorporation of hydrogen into a natural gas pipeline network is considered a method for delivering pure hydrogen to the market. To extract hydrogen from natural gas closer to the end-use point, various separation and purification technologies have been employed. Three techniques—pressure swing adsorption (PSA), membrane separation, and electrochemical hydrogen separation—can be utilized to extract hydrogen from hydrogen-natural gas blends. It is noted that blends containing less than 5–15% hydrogen (volume) typically pose minor issues, dependent on site-specific pipeline conditions and natural gas compositions. However, blending in the range of 15–50% hydrogen requires more substantial modifications, such as converting large household appliances or enhancing compression capacity along the distribution path for industrial users [39].

When transporting natural gas, approximately 0.3% of the pumped natural gas volume is consumed at compressor stations every 100–120 km to facilitate movement. To estimate energy costs for transporting hydrogen and natural gas through the same pipeline, accounting for the viscosities of hydrogen and methane at equal energy flows, let us analyze the power required for pumping N (W).

$$\mathrm{N} = \mathrm{V_o}\Delta \mathrm{p} = \frac{\pi}{4} D^2 v \Delta p = \frac{\pi}{4} D^2 v \frac{1}{2} \rho v^2 \xi \tag{1}$$

where V_o – volumetric flow, m^3/s; Δp – pressure drop, Pa; D – pipeline diameter, m; v – gas speed, m/s; ρ – gas density, kg/m^3; ξ – resistance coefficient; $Re = \rho v D/\mu$ – Reynolds number; n = 0.25 – for turbulent gas flow in the pipe; μ – dynamic viscosity, Pa•s.

Energy flow through the pipeline (J/kg)

$$\mathrm{Q} = \mathrm{V_o}\, \rho \mathrm{H_v} \tag{2}$$

where H_V is the highest calorific value of the transported gas.

From Eqs. (1), (2) we obtain the ratio of power (energy consumption) required for pumping hydrogen and methane Eq. (3):

$$\left(\frac{N_{H_2}}{N_{CH_4}}\right) = \left(\frac{N_{H_2}}{N_{CH_4}}\right)^n \left(\frac{\rho_{CH_4}}{\rho_{H_2}}\right)^2 \left(\frac{H_{vCH_4}}{H_{vH_2}}\right)^{3-n} \tag{3}$$

Due to the low density of hydrogen, the flow rate needs to be increased by about three times. While the increase in flow resistance is partially offset by viscosity differences, transmitting an equivalent amount of energy in the form of hydrogen through a pipeline demands approximately 4.6 times more energy than natural gas (**Figure** 7) [40]. During transportation, only 70–80% of the original hydrogen will be transferred over distances ranging from 2.5 to 4 thousand km.

Li et al. [41] conducted a numerical investigation on the Joule-Thompson (J-T) coefficient of natural gas at different hydrogen blending ratios, demonstrating a roughly linear decrease in the J-T coefficient of the natural gas-hydrogen mixture with increasing hydrogen blending ratio. Their findings also indicated a 40–50% reduction in the J-T coefficient when the hydrogen blending ratio reached 30% (mole fraction) compared to that of natural gas. Zhou et al. [42] reported on the hydrogen-blended gas-electricity integrated energy system, emphasizing the superiority of hydrogen blending in the upper line of the natural gas network over the lower line. They found that a concentrated hydrogen blending strategy is more effective than a dispersed one. Wu et al. [43] summarized recent research on the hydrogen-induced failure of high-strength pipeline steels in hydrogen-blended natural gas transmission. Zhang et al. [44] established a mathematical model for Hydrogen-Blended Natural Gas (HBNG)

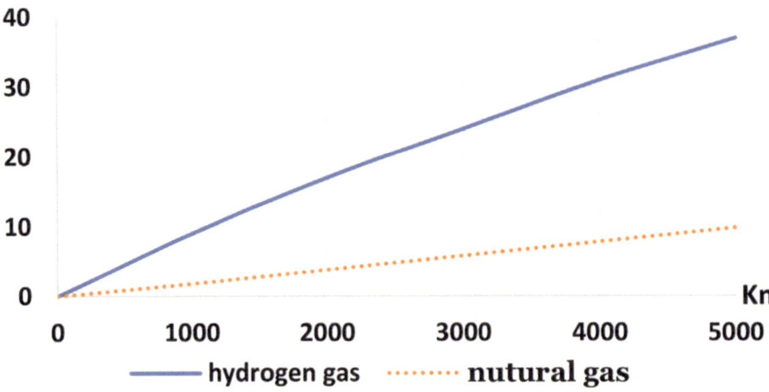

Figure 7.
Dependence of gas consumption for pumping on distance.

transportation, exploring the influences of hydrogen blending on hydraulic and thermal characteristics of natural gas pipelines and networks. Their results indicated that hydrogen blending could reduce pipeline friction resistance and increase volume flow rate. Additionally, they observed performance degradation of centrifugal compressors with increasing hydrogen blending ratio, leading to a shift in the operating point toward higher volume flow rates and lower pressure [44]. According to the European Naturally project [39], introducing hydrogen into the natural gas network holds potential advantage.

Utilizing the existing network of natural gas pipelines for hydrogen transport is a crucial aspect of the future hydrogen economy. Presently, the Unified Gas Supply System (UGSS) of JSC NC "QazaqGaz" possesses a significantly greater energy transmission capacity than power transmission networks and is fundamentally prepared to receive hydrogen and its mixtures with other flammable gases. The UGSS, the world's largest gas transportation system, is a unique technological complex encompassing gas production, processing, transportation, storage, and distribution facilities. It ensures a continuous gas supply cycle from the well to the end consumer. The group of companies of JSC "NC "QazaqGaz" operates gas pipelines with a total length of about 76 thousand km. Including 20 thousand km of main gas pipelines with an annual capacity of up to 267.8 billionm^3 and gas distribution networks with a length of about 56 thousand km, transportation gas is provided by 42 compressor stations and 238 gas injection units [45]. Consequently, most conditions necessary for hydrogen transportation have already been established. However, the UGSS is currently fully loaded, and the use of the existing gas pipeline network seems feasible only during the transition period to the hydrogen economy. To use a mixture of hydrogen and natural gas, creating cost-effective and efficient technologies for gas separation and hydrogen purification will be essential.

Experimental studies investigating the possibility of transporting hydrogen using steel pipelines designed for natural gas [41] revealed that hydrogen losses from the system are 3–3.5 times greater than the volume of natural gas losses. However, given that the heat of combustion of hydrogen is approximately three times greater, the energy losses are roughly equivalent. Notably, during the 6-month experiment, there were no instances of self-ignition due to hydrogen leakage through fittings, and the materials of the pipelines and seals remained unchanged.

Nevertheless, the potential hydrogen embrittlement of steel structures remains a focal point of concern [15, 30, 46]. The current natural gas infrastructure may not be suitable for transporting and distributing hydrogen due to the utilization of insufficient quality metals in these systems. To draw definitive conclusions about the suitability of existing gas transmission systems for pumping hydrogen, comprehensive research and development are necessary to study the materials of modern gas pipelines. This research is crucial, especially considering that energy transmission through a gas pipeline in the form of hydrogen over distances of 2–3 thousand km is 2–4 times more economical than energy transmission through power lines. Additionally, pumping hydrogen through pipeline transport offers the advantage of accumulating and storing hydrogen in underground and above-ground facilities under pressure, delivering it to consumers at the required time and quantity.

Despite the numerous advantages of blending hydrogen with natural gas, higher blend levels pose considerable challenges in terms of pipeline materials, safety considerations, and modifications required for end-use applications. Evaluating the substantial costs associated with accommodating elevated hydrogen blends in a specific pipeline system is crucial, and these expenses must be carefully weighed against the benefits of incorporating hydrogen as a component in natural gas blends. Beyond a 50% blend level, more complex issues arise in multiple aspects, including pipeline materials, safety concerns, and the adjustments needed for end-use appliances or other applications [25].

Despite the promising results offered by hydrogen transportation, significant efforts are required to address various safety issues, such as material embrittlement and container leakage. Challenges include its wide flammability range and the minimal energy needed for ignition. These safety concerns present potential barriers that must be effectively addressed to fully open up the hydrogen market.

2.5 By carriers

Physical transportation of hydrogen is usually done in various high-pressure and cryogenic tanks made of different types of materials, which should not interact with the hydrogen or perform any other reactions. We mentioned above that the traditional transportation techniques for hydrogen are high-pressure gas cylinders and liquid hydrogen that belong to the category of physical transportation. There is a possibility of material-based hydrogen transportation consisting of chemical and physical carriers. Using these types of transportation methods, the existing infrastructure would avoid many problems associated with the delivery of hydrogen in gaseous or liquid forms and reduce costs. In chemical carriers, hydrogen molecules are split into atoms and integrated with the chemical structure of the material. Among all, metal hydrides (for example, $LiAlH4$) are the most famous group of materials that can be used for chemical carriers [47]. Metal hydride carriers have the ability to absorb and desorb hydrogen at either room temperature or through heating of the tank. The main challenges of chemical carriers' materials are the cost, weight, and operating temperature, enhancing the charge-discharge rate and controlling the formation of unwanted gases during decomposition. Metal hydride carriers are also called "rechargeable" carriers; which are transported to a fuel station, where hydrogen is extracted from them, and then returned for a new refueling. Such carriers include, for example, metal hydrides. When using metal hydrides as carriers, it is advisable to use the same hydrides in the giving and receiving water system, then the heat released by the receiving system can be used to separate the water from the delivery system. A promising chemical carriers alternative is also LOHCs such as N-ethyl carbazole, methanol, dibenzyltoluene, toluene, and others, where the hydrogen is bonded chemically with hydrogen-lean molecules and is released through a catalytic dehydrogenation (Re. (4)) [48].

$$C_{10}H_{18} \rightleftarrows C_{10}H_8 + 5H_2 \tag{4}$$

These transportation options are attractive due to their easy manageability under ambient conditions, the transport and release processes do not emit CO_2, and the carrier liquid is not consumed and can be used repeatedly. These carriers are non-toxic and non-corrosive but have a low storage capacity which can limit their applications [49].

Alongside organic hydrogen carriers, ammonia is a compound where hydrogen is similarly bonded with a lean hydrogen molecule and released through dehydrogenation. Typically, hydrogen can be accumulated in ammonia through the conventional Haber-Bosch ammonia production process, which is responsible for around 85% of total worldwide ammonia production. The produced ammonia can then be transported through pipelines, tank cars, and tanker vessels. Thus, at normal temperature, ammonia liquefies at a pressure of 1.0 MPa and it can be transported through pipes and stored in liquid form (ammonia liquefaction temperature –239.76 K, critical temperature 405 K). After shipment, the hydrogen is released from ammonia through the catalytic decomposition process at a temperature of 527–627 K and atmospheric pressure (Re. (5)), which is a highly energy-intensive process. Therefore, efforts are necessary to improve their energy efficiency, reliability, and scalability [50].

$$NH_3 = 1/2\mathrm{N}_2 + 3/2\mathrm{H}_2 \tag{5}$$

To produce 1 kg of hydrogen, 5.65 kg of ammonia is required.

Another approach to material-based transportation involves using porous materials as physical carriers. The most promising options include Metal-Organic Frameworks (MOFs) and porous carbon materials, such as carbon nanotubes. This technique offers advantages such as a large surface area, low binding energy for hydrogen, quicker charging and discharging rates, and cost-effectiveness. However, challenges persist, including the weight of the carrier materials, the need for low temperatures and high pressures, and limitations in both gravimetric and volumetric hydrogen density. Despite these challenges, both methods could enhance safety during transportation by allowing for lower accumulation pressures and manageable properties. Nonetheless, they may not be suitable for high-demand scenarios and are typically transported via roads.

It appears that various methods of transporting hydrogen will be employed during its development as an energy source, with different degrees of utilization. These methods may be combined and utilized at different stages of market development, depending on how hydrogen is produced.

During the initial phase of transitioning to a hydrogen economy, trailers equipped with specialized containers under pressure could be utilized since the demand for hydrogen would likely be relatively low, and this approach minimizes hydrogen loss during transportation.

The advantages and disadvantages of the main methods of transporting hydrogen are reduced to **Table 1**. Using cryogenic tankers for hydrogen delivery proves to be

Transportation method	Advantages	Disadvantages
Hydrogen gas		
Pipeline transport	• Highest cost-effectiveness for large volumes of hydrogen • No thermodynamic limitations to reduce transportation costs • Low power consumption • Transportation safety • Environmentally friendly • Use of existing pipelines systems for natural gas and oil • Accumulation and storage in underground gas storage facilities under pressure and supply through gas pipelines to consumers at the right time in the right quantity	• Large investments in the construction of special pipelines • Very high transportation costs for small volumes • Complex and expensive procedure for obtaining permits for land acquisition, construction, etc. • The need for comprehensive R&D to study the hydrogen resistance of existing pipe steels, the features of underground gas storage facilities, the creation of new materials, fittings, compressors, etc.
Container transportation	• No hydrogen loss • No need to create storage infrastructure at the point of consumption	• Suitable only for small consumers • High cost of transportation
Liquid hydrogen		
Cryogenic tanks	• High energy density and small volume • Relative cheapness and efficiency of cryogenic tanks • Minimizing the need for compression at points of consumption	• High power consumption and high cost • Impossibility of reducing cost when long-term use • Difficulty in handling cryogenic liquids
Bound hydrogen		
Carriers	• Minimum cost of transportation in the future • Use of existing infrastructure • Moderate pressures and temperatures in the system delivery • Possibility of reducing storage costs	• Difficult to use on site due to the need for transformation for unloading • Increased energy consumption • Possibility of impurities entering hydrogen gas • Availability of idle range of the carrier for recharging The need to conduct comprehensive R&D, including ensuring safety and impact on the environment

Table 1.
Main methods of transporting hydrogen.

the most cost-effective for the average consumer, especially when transporting larger quantities of hydrogen compared to trailers with pressure containers, and it enables delivery to all geographical regions.

Pipeline systems are best suited for transporting hydrogen to areas with high demand, especially as more production facilities connect to the network. Economic considerations will always influence the preferred method of delivery. For instance, establishing gas distribution lines in urban areas may pose challenges. A typical delivery scenario might involve transmitting hydrogen via a pipeline from a central plant to a terminal, from which further delivery could occur via trailers, cryogenic tanks, or cargo transport vehicles.

When selecting hydrogen transportation methods, safety considerations must be taken into account. Significant risks include potential disruptions to power supplies for large populations due to technogenic disasters, systemic accidents, or deliberate acts such as terrorism. Therefore, employing cutting-edge technologies, such as durable pipeline materials and remote-controlled sensors, for constructing new underground pipelines is particularly pertinent. This approach would necessitate implementing pipeline patrol programs (potentially different from those for natural gas), safety regulations for excavation, and other measures. Transporting hydrogen through underground pipelines is preferable in terms of ensuring the safety of the population, especially in the face of potential terrorist threats.

3. Hydrogen usage

Hydrogen, the most abundant and simplest element in the universe, is primarily found on Earth in compounds form with other elements. For instance, it combines with oxygen to create water (H_2O) and with carbon to form hydrocarbons, which are present in fossil fuels and various other resources. Although hydrogen has been utilized in chemical and industrial applications for over a century, recent investments by both markets and governments in hydrogen as an energy source have sparked increased interest in hydrogen production. When hydrogen is combusted, it mainly emits water vapor, making it a crucial component in efforts to reduce greenhouse gas emissions. Additionally, hydrogen is seen as a key solution for storing energy generated from conventional sources such as renewable energy, natural gas, and nuclear power. The International Energy Agency's (IEF) 2023 Global Hydrogen Outlook (GBO – 2023) reported that the Global hydrogen use reached 95 Mt. in 2022, and which is nearly 3% increased from their revised estimate for 2021 [51]. Hydrogen use has grown mainly in all countries except Europe. The only reason for the decline in the use of hydrogen in Europe was attributed to the Russian invasion of Ukraine because as a result of the war, European chemical plants reduced production volumes, which led to a 6% reduction in the use of hydrogen in Europe. However, there was a strong growth of 7% in North America and the Middle East. China remains the world's largest consumer of hydrogen with a 0.5% increase in hydrogen use (**Figure 8**). Areas of wide hydrogen uses are briefly as follows:

3.1 Oil refining

Hydrogen plays a crucial role in the refining of petroleum, aiding in the desulfurization and catalytic cracking of long-chain hydrocarbons. Approximately one-quarter of global production is dedicated to converting low-grade crude oils (particularly from tar sands) into high-energy transport fuels like gasoline and diesel. The process involves converting heavy aromatic feedstock into lighter alkane hydrocarbon products under intense pressures (7000–14,000 kPa) and high temperatures (400–800°C), using hydrogen and specialized catalysts [52]. Moreover, hydrogen is essential for eliminating impurities such as sulfur from these fuels. In 2022, the use of hydrogen in oil refining exceeded 41 million metric tons, surpassing the previous peak in 2018. The most significant rise in year-over-year demand originated from

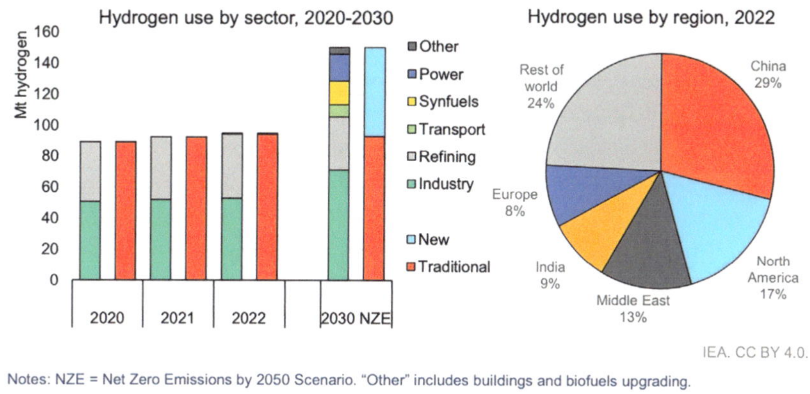

Figure 8.
Usages of hydrogen across various sectors and regions, both historically and within the context of the net zero emissions by 2050 scenario, 2020–2030 [51].

North America and the Middle East, collectively representing over 1 million metric tons, or approximately 75% of global growth in 2022 [51].

3.2 Production of ammonia

The production of ammonia through the synthesis of hydrogen and nitrogen, accounting for approximately 180 million metric tons per year or 1 petawatt-hour, constitutes more than half of the global demand for pure hydrogen. Of the 53 Mt. of hydrogen used in industry in 2022, about 60% was for ammonia production [53]. This process relies on the Haber-Bosch method (Re. (6)). Primarily, ammonia is utilized for agricultural fertilizers, with a portion being employed, in the form of ammonium nitrate combined with diesel fuel, for mining explosives. Additionally, it can be utilized as a transportation fuel or subjected to cracking to yield hydrogen for fuel purposes. The Haber process uses 3–5% of the world's natural gas to produce the hydrogen, and the nitrogen is extracted from the air by cooling it [54]. Looking ahead, ammonia could play a significant role in hydrogen storage and transportation, as discussed in upper sections. It also holds potential as a fuel source. In Japan, initiatives are underway to explore the co-firing of ammonia with coal in boilers and with natural gas in combustion turbines [55]. Moreover, ammonia shows promise as a maritime fuel, requiring only minor modifications for use in ship engines, and it can also be utilized in certain fuel cell technologies.

$$N_2 + 3H_2 \rightleftharpoons 2NH_3 \ \Delta H - 92 \text{ kJ/mol} \tag{6}$$

3.3 Production of methanol and DME fuels

Given the problems of storage and transportation of hydrogen itself, as well as the radical change of fuel cell cars, methanol (CH_3OH) can be obtained by reacting hydrogen gas with atmospheric CO_2 gas (Re. (7)) [56]. Methanol has several own potentials. First, dimethyl ether (DME) can be obtained from methanol, which is made by dehydrating several methanol molecules (Re. (8)). It is a gas but can be stored under low pressure as a liquid. Second, methanol is preferred for gasoline engines, dimethyl ether (CH_3-O-CH_3) for diesel engines [57]. Methanol and DME production is at relatively low temperature. Third, the energy density of methanol and DME is 16 MJ/L and 18–19 MJ/L, respectively, which is lower than petroleum-based fuel, but usable and easy to store.

$$CO_2 + 3H_2 \Rightarrow CH_3OH + H_2O \quad \Delta H - 49.5\,kJ/mol\ at\ 25^{\circ}C \tag{7}$$

$$2CH_3OH \Rightarrow CH_3OCH_3 + H_2OH23\,kJ/mol \tag{8}$$

3.4 Liquid hydrocarbon fuels

Coal and biomass have long served as the foundation for liquid hydrocarbon fuels, relying on hydrogen for their conversion [58]. Originating in 1920s Germany, the Fischer-Tropsch process, primarily coal-based, fueled a significant portion of Germany's World War II efforts and later became instrumental in South Africa's oil production, notably by Sasol. This process, requiring substantial hydrogen, catalyzes carbon monoxide to yield liquid hydrocarbons, now facilitated by coal gasification. Approximately 14,600 tons of coal yield 25,000 barrels of synfuel "oil", alongside 25,000 tons of CO_2. Nuclear power offers avenues for enhancement: nuclear hydrogen sources coupled with process heat could boost hydrocarbon output and slash CO_2 emissions, while a hybrid system utilizes nuclear electricity for water electrolysis, yielding hydrogen for coal gasification. Conversely, biomass undergoes hydrotreating or Fischer-Tropsch processing to produce liquid biofuels, demonstrating sustainable alternatives in liquid fuel production [59].

3.5 Fuel cells

Hydrogen is predominantly utilized in fuel cell electric vehicles (FCEVs) for transportation purposes. Unlike conventional batteries, fuel cells generate electricity through a chemical reaction using external hydrogen fuel and oxygen from the air. Proton exchange membrane (PEM) fuel cells, the primary type used in cars and heavy vehicles, operate at temperatures of around 80–90°C [60]. They offer high volumetric power density and long life but require high-purity hydrogen and costly noble metal catalysts, typically platinum. Although they theoretically achieve about 60% efficiency in converting chemical energy to electrical energy, practical efficiency is approximately half of that. Alternatively, alkaline fuel cells (AFCs) operate at around 200°C, boasting efficiency above 60%. Developed since the 1960s, AFCs have been employed by NASA in space missions due to their reliability [61]. They are cost-effective, utilizing non-noble metal catalysts and tolerating less-pure hydrogen from ammonia cracking. However, commercialization is limited by CO_2 poisoning, which leads to insoluble carbonate formation.

3.6 Reductant for metallurgy

Metallurgical coke, primarily carbon, plays a crucial role in steelmaking as a reductant, yet advancements in utilizing natural gas for direct iron reduction are emerging to mitigate CO_2 emissions [62]. Currently, blast furnaces, fueled by coke, dominate steel production, while electric arc furnaces (EAFs) and direct-reduced iron (DRI) methods are gaining traction. EAFs, predominantly powered by electricity, offer synergy with nuclear energy, while DRI, facilitated by hydrogen, presents an avenue for clean production. The steel sector accounts for a substantial portion of global hydrogen usage, with green steel initiatives, such as hydrogen in Europe's vision and projects like HYBRIT in Sweden, driving innovation. Plans for green steel production in Australia and Russia underscore the industry's shift towards sustainable practices [63], utilizing electrolysis and renewable energy sources to meet hydrogen demands and reduce carbon footprint.

4. Conclusion

In the realm of transportation, transporting hydrogen via tube trailers without the need for liquefaction proves to be more economically efficient, resulting in a savings of $2.86 per kilogram of delivered H_2 in smaller-scale power plants. The issue of hydrogen boil-off during transport stands out as a critical concern. The extent of embrittlement largely hinges on factors such as the material composition of the pipeline and the prevailing temperature and pressure conditions. Furthermore, the embrittlement of pipeline steels and other construction materials can lead to deterioration in mechanical properties and the development of cracks, ultimately resulting in pipeline failures. Additionally, it is observed that blending hydrogen into the upper line of the natural gas network yields superior results compared to blending it into the lower line. Moreover, a strategy focusing on concentrated hydrogen blending outperforms a dispersed approach.

Hydrogen stands as a pivotal element in various industrial processes and emerging technologies, playing a vital role in sectors ranging from oil refining to metallurgy and transports. Its versatility as an energy carrier enables its use in the desulfurization and catalytic cracking of petroleum, as well as in the production of essential chemicals like ammonia and methanol. Moreover, hydrogen facilitates the synthesis of liquid hydrocarbon fuels from coal and biomass, offering sustainable alternatives for the future. In the realm of transportation, hydrogen fuel cells present an efficient and clean solution, though challenges remain in terms of infrastructure and cost-effectiveness. Furthermore, hydrogen's significance in metallurgy, particularly in steel production, highlights its potential to drive the transition toward greener industrial practices. With ongoing advancements and initiatives aimed at harnessing its potential, hydrogen emerges as a key player in the pursuit of a more sustainable and low-carbon future. Consequently, these challenges have the potential to disrupt the distribution and delivery of hydrogen. Therefore, the successful realization of a hydrogen-based economy is contingent upon the ability to identify the most promising areas for future advancements in hydrogen transportation, usage, production, and storage.

Author details

Akbar Dauletbay
Farabi University, Almaty, Kazakhstan

*Address all correspondence to: akbar.dauletbay@kaznu.kz

References

[1] Faye O, Szpunar JA. An efficient way to suppress the competition between adsorption of H_2 and desorption of *n* H_2 –Nb complex from graphene sheet: A promising approach to H_2 storage. Journal of Physical Chemistry C. 2018;**122**(50):28506-28517

[2] Faye O, Hussain T, Karton A, Szpunar J. Tailoring the capability of carbon nitride (C_3N) nanosheets toward hydrogen storage upon light transition metal decoration. Nanotechnology. 2019;**30**(7):075404

[3] Faye O, Szpunar JA, Szpunar B, Beye AC. Hydrogen adsorption and storage on palladium – Functionalized graphene with NH-dopant: A first principles calculation. Applied Surface Science. 2017;**392**:362-374

[4] Faye O, Eduok U, Szpunar J, Szpunar B, Samoura A, Beye A. Hydrogen storage on bare Cu atom and Cu-functionalized boron-doped graphene: A first principles study. Chicago International Journal of Hydrogen Energy. 2017;**42**(7):4233-4243

[5] Miller EL, Papageorgopoulos D, Stetson N, Randolph K, Peterson D, Cierpik-Gold K, et al. U.S. Department of Energy hydrogen and fuel cells program: Progress, challenges and future directions. MRS Advances. 2016;**1**(42):2839-2855

[6] RENA. Renewable Power Generation Costs in 2017, International Renewable Energy Agency, Abu Dhabi. 2018. Available from: https://www.irena.org/-/media/Files/IRENA/Agency/Publication/2018/Jan/IRENA_2017_Power_Costs_2018.pdf?rev=05d2a50cb37847f088d1e95db9c67921

[7] Ji M, Wang J. Review and comparison of various hydrogen production methods based on costs and life cycle impact assessment indicators. International Journal of Hydrogen Energy. 2021;**46**(78):38612-38635

[8] Ma Y, Wang XR, Li T, Zhang J, Gao J, Sun ZY. Hydrogen and ethanol: Production, storage, and transportation. International Journal of Hydrogen Energy. 2021;**46**(54):27330-27348

[9] Hosseini SE, Wahid MA. Hydrogen production from renewable and sustainable energy resources: Promising green energy carrier for clean development. Renewable and Sustainable Energy Reviews. 2016;**57**:850-866

[10] Hirscher M, Yartys VA, Baricco M, Bellosta von Colbe J, Blanchard D, Bowman RC, et al. Materials for hydrogen-based energy storage – Past, recent progress and future outlook. Journal of Alloys and Compounds. 2020;**827**:153548

[11] Lee J, W. Hydrogen, Properties, Uses, & Facts. Chicago: Britannica [Internet]. 2024. Available from: https://www.britannica.com/science/hydrogen; [Accessed: February 18, 2024]

[12] Savitri D. 18 Uses of Hydrogen — Commercial, and Miscellaneous. Bangalore, Karnataka, India: Techiescientist [Internet]. 17 Oct 2022. Available from: https://techiescientist.com/uses-of-hydrogen/ [Accessed: February 18, 2024]

[13] Hydrogen Uses - Hydrogen Portal [Internet]. Available from: https://hydrogen-portal.com/technologies/hydrogen-uses/ [Accessed: February 18, 2024]

[14] Sunita S. Hydrogen: A Clean, Flexible Energy Carrier. Washington,

DC: Department of Energy, Office of Energy Efficiency & Renewable Energy [Internet]. 2017. Available from: https://www.energy.gov/eere/articles/hydrogen-clean-flexible-energy-carrier; [Accessed: February 18, 2024]

[15] Züttel A. Materials for hydrogen storage. Materials Today. 2003;**6**(9):24-33

[16] Ogden JM. Hydrogen energy systems studies. Process System Engineering. 2014;**5-7**(October):125-157

[17] Гамбург ДЮ, Семенов ВП, Дубовкин НФ, Смирнова ЛМ. Водород. Свойства, получение, хранение, транспортирование, применение. Справочник. Москва: Химия; 1989:1-672

[18] Freedom CAR & Fuel Partnership. Hydrogen Delivery Technologies Roadmap [Internet]. Washington DC: Freedom CAR & Fuel Partnership; 2005. Available from: https://www1.eere.energy.gov/vehiclesandfuels/pdfs/program/delivery_tech_team_roadmap.pdf

[19] Wurster R, Zittel W. Hydrogen Energy, Workshop Energy Technologies to Reduce CO_2 Emission in Europe [Petten]. Petten (Nederland): Energieonderzoek Centrum Nederland [Internet]. 1994. Available from: https://scholar.google.com/scholar_lookup?&title=WorkshoponEnergyTechnologiestoReduceCO2emissionsinEurope%3AProspects%2CCompetition%2CSynergy&publication_year=1994&author=Wurster%2CR.&author=Zittel%2CW [Accessed: February 18, 2024]

[20] Smith B, Ridge National Laboratory O. Fiber-Reinforced Polymer Pipelines for Hydrogen Delivery. Excerpt from 2007 DOE Hydrogen Program Annual Progress Report; San Antonio, Texas; 2007

[21] Corrosion Resistant Fiberglass Piping Systems For Oil & Gas Applications

[22] Mintz M, Ringer M, Pastor M. H2A Delivery Analysis. Excerpt from DOE Hydrogen Program 2006 Annual Progress Report; Toronto. 6 June 2006

[23] Reddi K, Mintz M, Elgowainy A, Sutherland E. Building a hydrogen infrastructure in the United States. In: Compendium of Hydrogen Energy. Sawston, Cambridge: Woodhead Publishing; 2016. pp. 293-319

[24] Demir ME, Dincer I. Cost assessment and evaluation of various hydrogen delivery scenarios. International Journal of Hydrogen Energy. 2018;**43**(22):10420-10430

[25] Melaina MW, Antonia O, Penev M. Blending Hydrogen into Natural Gas Pipeline Networks: A Review of Key Issues. Golden, CO (United States): National Renewable Energy Laborator; 2013

[26] Europe: hydrogen pipeline length by country. 2023. Statista [Internet]. Available from: https://www.statista.com/statistics/1389658/proposed-hydrogen-pipeline-infrastructure-length-in-europe-by-country/ [Accessed: February 18, 2024]

[27] 1,000 km hydrogen pipelines are under construction in China [Internet]. Available from: https://chinahydrogen.substack.com/p/1000-km-hydrogen-pipelines-are-under [Accessed: February 18, 2024]

[28] Which countries are building hydrogen pipelines fastest? World Economic Forum [Internet]. Available from: https://www.weforum.org/agenda/2023/12/hydrogen-pipelines-countries-fastest/ [Accessed: February 18, 2024]

[29] Home. FIBA Canning Inc. Tank Truck and Tube Trailer Repair [Internet]. Available from: https://www.fibacanning.com/ [Accessed: February 18, 2024]

[30] TransPower - Hydrogen Delivery Systems [Internet]. Available from: https://transpowerusa.com/zero_emission_solutions/hydrogen_delivery_systems/ [Accessed: February 18, 2024]

[31] Ares JR, Leardini F, Díaz-Chao P, Bodega J, Fernández JF, Ferrer IJ, et al. Ultrasonic irradiation as a tool to modify the H-desorption from hydrides: MgH_2 suspended in decane. Ultrasonics Sonochemistry. 2009;**16**(6):810-816

[32] Kimura T, Fujita M, Sohmiya H, Ando T. Ultrasonic acceleration of iodination of unactivated aliphatic hydrocarbons. Ultrasonics Sonochemistry. 2002;**9**(4):205-207

[33] Chave T, Nikitenko SI, Granier D, Zemb T. Sonochemical reactions with mesoporous alumina. Ultrasonics Sonochemistry. 2009;**16**(4):481-487

[34] Hiroi S, Hosokai S, Akiyama T. Ultrasonic irradiation on hydrolysis of magnesium hydride to enhance hydrogen generation. International Journal of Hydrogen Energy. 2011;**36**(2):1442-1447

[35] Merouani S, Hamdaoui O, Rezgui Y, Guemini M. Computational engineering study of hydrogen production via ultrasonic cavitation in water. International Journal of Hydrogen Energy. 2016;**41**(2):832-844

[36] Черемных ОЯ. Creation, design improvement, development prospect of vehicles for liquid hydrogen. Engineering Journal: Science and Innovation. 2017;**3**:64

[37] Hydrogen Facilities [Internet]. Available from: https://www.cryogenmash.ru/en/catalog/hydrogen-facilities/ [Accessed: February 18, 2024]

[38] Schlapbach L, Züttel A. Hydrogen-storage materials for mobile applications. Nature. 2001;**414**(6861):353-358

[39] Reitenbach V, Ganzer L, Albrecht D, Hagemann B. Influence of added hydrogen on underground gas storage: A review of key issues. Environment and Earth Science. 2015;**73**(11):6927-6937

[40] Bossel U, Eliasson B, Taylor G. The future of the hydrogen economy: Bright or bleak? Cogeneration & Distributed Generation Journal. 2003;**18**(3):29-70

[41] Li J, Su Y, Yu B, Wang P, Sun D. Influences of hydrogen blending on the Joule–Thomson coefficient of natural gas. ACS Omega. 2021;**6**(26):16722-16735

[42] Zhou D, Yan S, Huang D, Shao T, Xiao W, Hao J, et al. Modeling and simulation of the hydrogen blended gas-electricity integrated energy system and influence analysis of hydrogen blending modes. Energy. 2022;**239**:121629

[43] Wu X, Zhang H, Yang M, Jia W, Qiu Y, Lan L. From the perspective of new technology of blending hydrogen into natural gas pipelines transmission: Mechanism, experimental study, and suggestions for further work of hydrogen embrittlement in high-strength pipeline steels. International Journal of Hydrogen Energy. 2022;**47**(12):8071-8090

[44] Zhang H, Li J, Su Y, Wang P, Yu B. Effects of hydrogen blending on hydraulic and thermal characteristics of natural gas pipeline and pipe network. Oil & Gas Science and Technology - Revue d'IFP Energies Nouvelles. 2021;**76**:70

[45] AO HK QazaqGaz - General information [Internet]. Available from: https://qazaqgaz.kz/en/obshchaya-informaciya [Accesed: February 18, 2024]

[46] Mintz M, National LA. Hydrogen Delivery Infrastructure Analysis. DOE Hydrogen Program FY 2010 Annual Progress Report. Washington, D.C: 2011

[47] Ren J, Musyoka NM, Langmi HW, Mathe M, Liao S. Current research trends and perspectives on materials-based hydrogen storage solutions: A critical review. International Journal of Hydrogen Energy. 2017;**42**(1):289-311

[48] Preuster P, Papp C, Wasserscheid P. Liquid organic hydrogen carriers (LOHCs): Toward a hydrogen-free hydrogen economy. Accounts of Chemical Research. 2017;**50**(1):74-85

[49] Teichmann D, Arlt W, Wasserscheid P. Liquid organic hydrogen carriers as an efficient vector for the transport and storage of renewable energy. International Journal of Hydrogen Energy. 2012;**37**(23):18118-18132

[50] Aziz M, Wijayanta AT, Nandiyanto ABD. Ammonia as effective hydrogen storage: A review on production, storage and utilization. Energies. 2020;**13**(12):3062

[51] IEA. Global Hydrogen Review 2023 – Analysis. France: IEA [Internet]; 2023. Available from: https://www.iea.org/reports/global-hydrogen-review-2023 [Accessed: February 18, 2024]

[52] Moradpoor I, Syri S, Santasalo-Aarnio A. Green hydrogen production for oil refining – Finnish case. Renewable and Sustainable Energy Reviews. 2023;**175**:113159

[53] Al-Breiki M, Bicer Y. Comparative cost assessment of sustainable energy carriers produced from natural gas accounting for boil-off gas and social cost of carbon. Energy Reports. 2020;**6**:1897-1909

[54] Manigandan S, Ryu JI, Praveen Kumar TR, Elgendi M. Hydrogen and ammonia as a primary fuel – A critical review of production technologies, diesel engine applications, and challenges. Fuel. 2023;**352**:129100

[55] JERA. Plan a for Decarbonization: Why Is Japan Using Ammonia for Thermal Power Generation? Discover JERA, JERA [Internet]. Japan; 2023. Available from: https://www.jera.co.jp/en/action/discover/025; [Accessed: February 18, 2024]

[56] Methanol fuel - Wikipedia [Internet]. Available from: https://en.wikipedia.org/wiki/Methanol_fuel#cite_note-3 [Accessed: February 18, 2024]

[57] Alternative Fuels Data Center: Dimethyl Ether [Internet]. Available from: https://afdc.energy.gov/fuels/emerging_dme.html [Accessed: February 18, 2024]

[58] Katzer JR. Coal and Biomass to Liquid Fuels. New York, NY: Springer; 2011. pp. 89-128

[59] Biomass explained - U.S. Energy Information Administration (EIA) [Internet]. Available from: https://www.eia.gov/energyexplained/biomass/ [Accessed: February 18, 2024]

[60] Alaswad A, Palumbo A, Dassisti M, Olabi AG. Fuel cell technologies, applications, and state of the art. A reference guide. In: Reference Module in Materials Science and Materials Engineering. Amsterdam: Elsevier; 2016

[61] Luo Y, Shi Y, Cai N. Bridging a bi-directional connection between electricity and fuels in hybrid multienergy systems. In: Hybrid Systems and Multi-Energy Networks for the Future Energy Internet. London, United Kingdom: Elsevier; 2021. pp. 41-84

[62] Sun M, Pang K, Barati M, Meng X. Hydrogen-based reduction technologies in low-carbon sustainable ironmaking and steelmaking: A review. Journal of Sustainable Metallurgy. 2023;**10**:10-25

[63] PCI coal for steelmaking the first metallurgical coal grade to be impacted by decarbonisation. IEEFA [Internet]. Available from: https://ieefa.org/resources/pci-coal-steelmaking-first-metallurgical-coal-grade-be-impacted-decarbonisation [Accessed: February 18, 2024]